P9-EMQ-006

trees

A VISUAL GUIDE

Tony Rodd and Jennifer Stackhouse

trees

A VISUAL GUIDE

UNIVERSITY OF CALIFORNIA PRESS
BERKELEY LOS ANGELES

University of California Press, one of the most distinguished university presses in the United States, enriches lives around the world by advancing scholarship in the humanities, social sciences, and natural sciences. Its activities are supported by the UC Press Foundation and by philanthropic contributions from individuals and institutions. For more information, visit www.ucpress.edu.

University of California Press
Berkeley and Los Angeles, California

Published by arrangement with
Weldon Owen Pty Ltd
61 Victoria Street, McMahons Point
Sydney, NSW 2060, Australia

Group Chief Executive Officer John Owen
President and Chief Executive Officer Terry Newell
Publisher Sheena Coupe
Creative Director Sue Burk
Vice President, International Sales Stuart Laurence
Vice President, Sales and New Business Development Amy Kaneko
Vice President Sales, Asia and Latin America Dawn Low
Administrator, International Sales Kristine Ravn

Managing Editor Jennifer Losco
Project Editor Ariana Klepac
Publishing Coordinator Mike Crowton
Designer Mark Thacker/Big Cat Design
Picture Researcher Joanna Collard
Art Buyer Trucie Henderson
Illustrator Peter Bull Art Studio
Information Graphics Andrew Davies/Creative Communication

Cataloging-in-Publication data for this title is on file with the Library of Congress.

ISBN 978-0-520-25650-7

Color reproduction by Chroma Graphics (Overseas) Pte Ltd
Printed by SNP Leefung Printers Ltd
Printed in China

A WELDON OWEN PRODUCTION

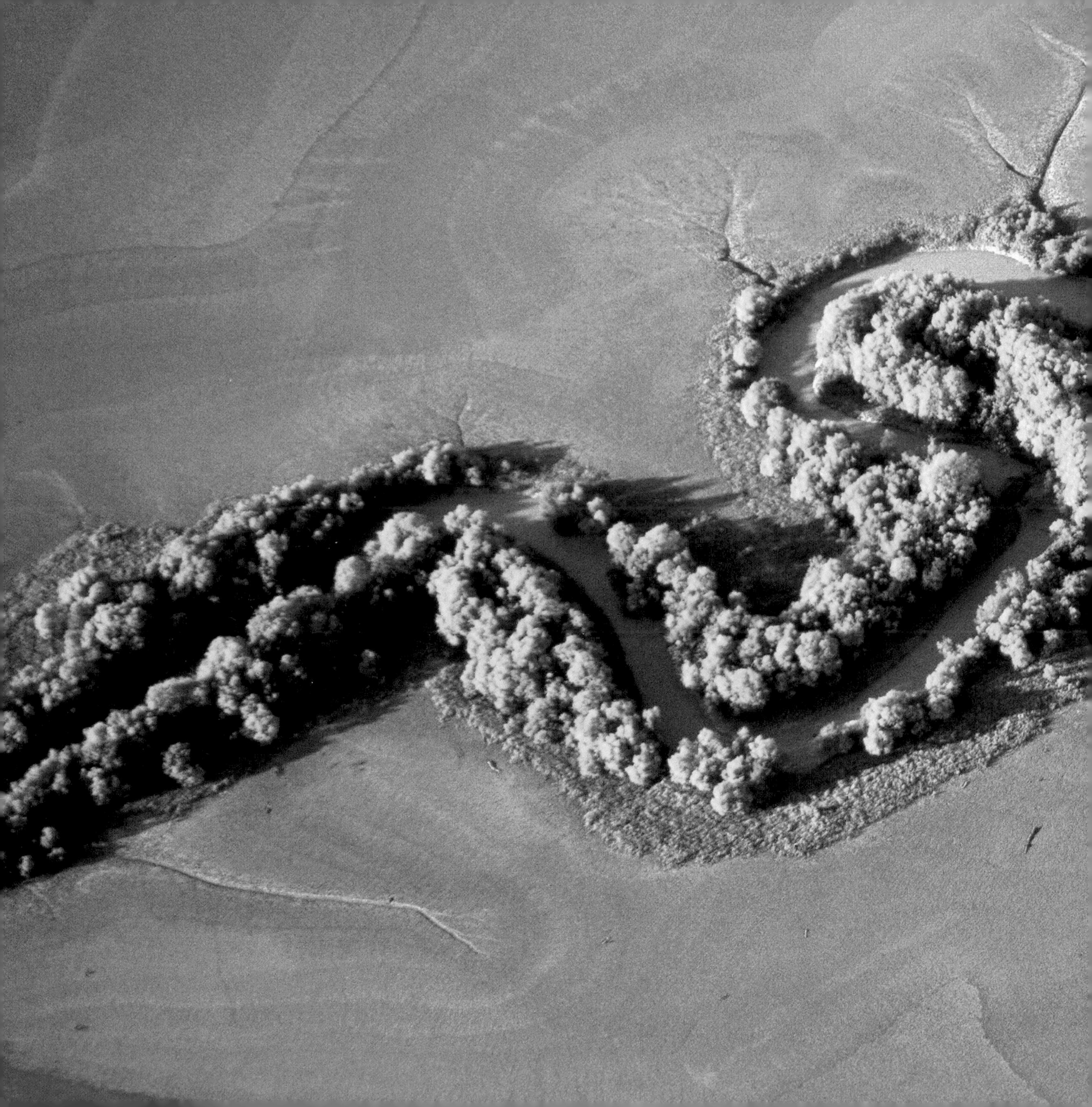

Contents

13 Introduction

14 A world of trees

88 Diversity and design

30 Form and function

176 Communities of life

220 Trees and the human world

262 An indispensable resource

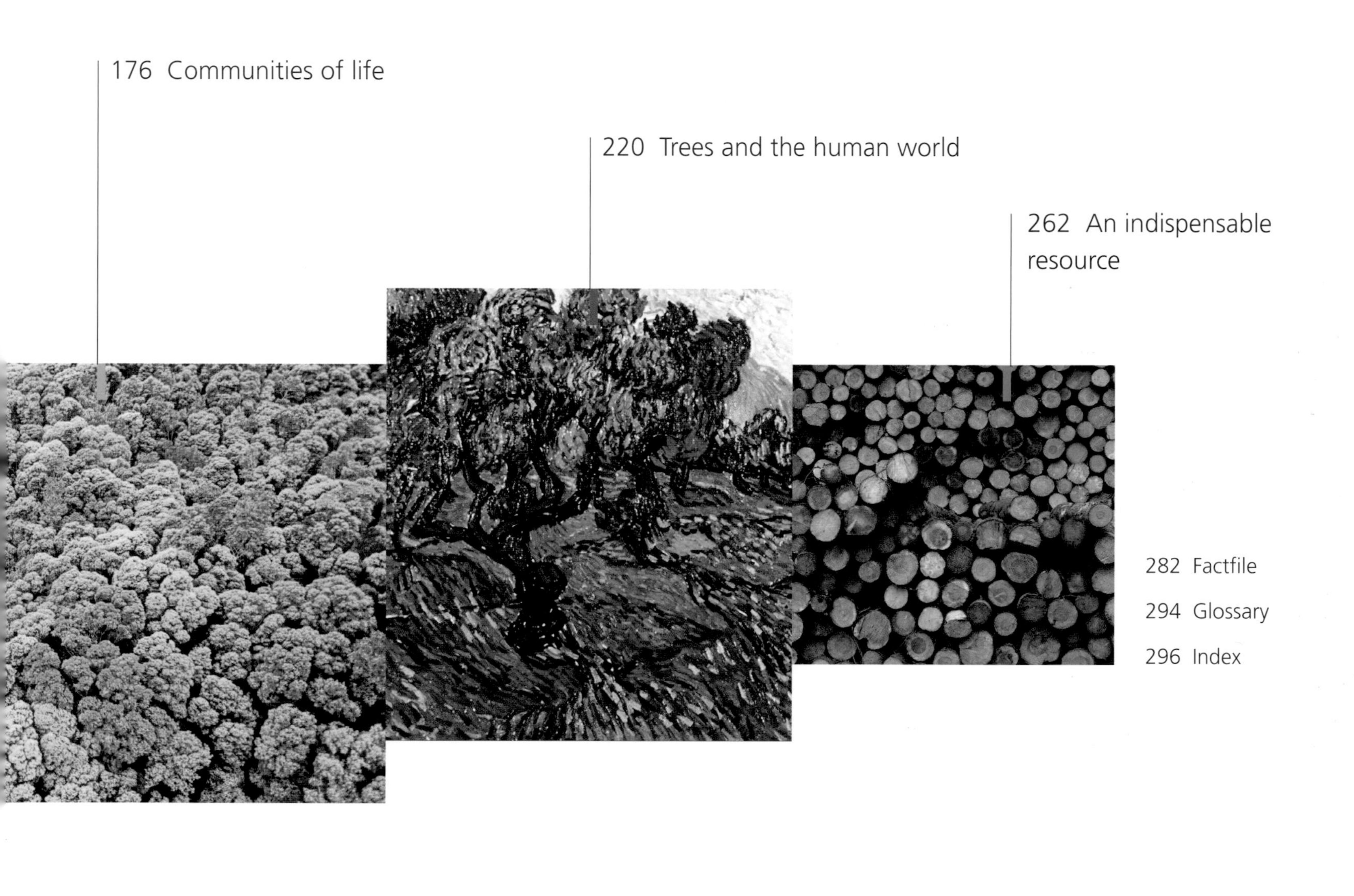

282 Factfile

294 Glossary

296 Index

Introduction

Trees are the "big game," or the "whales" of the plant world. But unlike those massive animals, trees are all around us and, without doubt, include in their number the world's largest living organisms. In fact, in the world's forests there are probably still growing millions of trees of mass greater than the largest whales (about 180 tons). We tend to take for granted the trees in our local environment. But even the most ordinary tree, whether growing in the street or in a city park, is worthy of contemplation. Stand below it and try to estimate its weight, height, spread of canopy, pattern of branching from trunk to limbs to successively smaller branches and twigs and finally the leaves. Think of how many leaves it has; what might be the total area of leaf surface, and how much light this intercepts—which determines how much of the sun's energy the tree can absorb and use in photosynthesis, as well as shading the ground beneath (or our house) from that same amount of solar energy. Consider the lifespan of any one leaf. And consider the root system—continually drawing water and minerals from the soil (except in winter or drought dormancy) and pumping them through the sapstream to the leaves, which return that water to the atmosphere as well as absorbing carbon-dioxide and releasing oxygen. This book will visit these and many other aspects of the tree world—the way a tree grows and functions; the varied architectures of trees; the diversity of trees in terms of species, genera and families; the major communities of trees around the world; and all the varied ways in which humans make use of trees, from timber and pharmaceuticals to shade and shelter.

A world of trees

A world of trees

Trees first appeared on Earth around 300 million years ago during the Carboniferous period, as giant horsetails, tree ferns and lycopods. Today trees exist in almost every climate, in a bewildering array of forms, some growing to around 380 feet (116 m), and others living for more than a thousand years.

What is a tree? 18

The large and the small 20

The old and the new 22

The art of naming trees 24

Climate and trees 26

What is a tree?

The word "tree" is not an easy one to define. Its central meaning (as typified by, say, a huge oak) is readily agreed to by all speakers of the language. Most of us agree on the essential features of a tree. A tree is typically tall: it is significantly taller than a tall person, maybe two or three times at very least. A tree is perennial: it should be able to continue growing and increasing in size for an indefinite number of years. A tree grows from the top: plants that put up new, short-lived stems from the base at regular intervals—for example banana plants (*Musa* spp.)—are not trees. A tree is (usually) woody: trees generally have strong, rigid trunks supporting their foliage. But true wood, laid down by a cylinder of cambium cells inside the bark, is found only in dicotyledons, conifers and ginkgos. What about cycads, palms, yuccas, cacti, tree ferns and other such tall, long-lived plants? They do not have true wood, but they fit into our concept of "tree" in most other ways. A tree can be single or multi-stemmed: if the height of a woody perennial plant is between about 10 feet (3 m) and 20 feet (6 m), we tend to call it a shrub rather than a tree if it branches from ground level into many stems. If a plant is under 10 feet (3 m) we would probably call it a shrub, and over 20 feet (6 m) a tree.

CLASSIFYING A PLANT AS A TREE

The botanical classification of a plant may be a poor guide to whether it is a tree or has some other growth form—such as shrub, herb or vine. Nearly all larger families of plants include a wide spectrum of growth forms, though in some families trees predominate, and in others, herbs (herb used here in the botanical sense of a non-woody plant). And many of the larger genera of plants, for example *Euphorbia*, include a range of forms from trees to tiny herbs. Even a single species or subspecies may grow into a tree, shrub or more rarely a vine, depending on the conditions it grows under and length of life they allow it to achieve. In fact this sort of variation is quite common among plants found in exposed mountain or coastal areas. In these situations there may be selection for a more shrubby, spreading growth-habit that evades strong winds, resulting in genetically fixed races of the species with those characteristics.

↓ **"Tree" is quite an arbitrary term.** Trees belong to many different orders and families of plants and can vary drastically in appearance, from the typical tree with tall trunk and spreading branches (such as the beech pictured), to tree ferns, cycads, cacti and succulents, palms and aquatic plants.

↑ **The Saguaro** is a tree in most senses, though it is also a cactus and a stem succulent—leafless with a fleshy, water-storing stem.

← **This elm tree** matches most people's concept of a typical tree—it is large, long-lived, has a single thick trunk and all its growth is from the top.

↓ **Classed as a pachycaul,** the Boojum Tree has a trunk that is swollen at the base and clothed in numerous short, thornlike lateral branches.

The large and the small

Trees come in many sizes, and variations in size are the result of differences in genetic potential, environment or a combination of both. An apple tree can never reach a fraction of the height of a redwood regardless of its growing conditions. And a Scots Pine growing on a craggy mountainside in the Scottish Highlands will not reach half the height of a Scots Pine in the deep, moist soil of an Austrian valley bottom. Of course when talking about trees, we should realize that the smaller size limits of many tree species are hardly trees at all, but shrubs. This is especially true for individuals growing on exposed ocean headlands, or on storm-swept mountains at the treeline. Turning to the tallest trees, there seems to be an upper limit to their height of somewhere around 400 feet (122 m), though heights approaching this seem only to have been reached by several of the conifers (*Sequoia*, *Pseudotsuga* and *Picea*) of North America's Pacific coast "rainbelt," and by *Eucalyptus regnans* in a small part of south-eastern Australia. Several of these species reportedly once had trees over 400 feet, but these were soon felled by pioneering timber-getters.

↓ **Below left to right:** Coast Redwood (*Sequoia sempervirens*), Umbrella Thorn (*Acacia tortilis*), Alerce (*Fitzroya cupressoides*) and Boojum Tree (*Fouquieria columnaris*). The size and shape of trees varies dramatically, from tall and narrow, short and spreading to tall and spreading. An adult human figure is shown next to each tree to indicate scale.

↑ **El Arbol del Tule** is a spectactular specimen of Montezuma Cypress (*Taxodium mucronatum*), with a trunk diameter of almost 40 feet (12 m).

↗ **The Grass Tree (*Xanthrrhoea glauca*)** is an evergreen swordleaf endemic to Australia that grows to a maximum height of 25 feet (8 m).

← **The massive Coast Redwoods (*Sequoia sempervirens*)** are the world's tallest trees.

↓ **Japanese Maple (*Acer palmatum*)** grows to approximately 40 feet (12 m) but has a graceful spreading habit that is much admired and is the reason why it is widely planted as an ornamental.

SMALLER TREES

Unlike the tallest, trees at the smaller limits of what we would call a tree occur all around the world in a great diversity of genera and species. In more arid habitats they are almost the norm. Semi-desert woodland or savanna often has a canopy height of between about 15 and 25 feet (4.5 and 7.5 m) and thousands of different trees grow in this vegetation in warmer parts of Africa, the Americas and Australia. Numerous *Acacia* species from these large regions seldom exceed about 25 feet (7.5 m) in height. In temperate regions we have the pinyon-juniper woodlands on the dry plateaus of Colorado and Arizona, where an abundant species such as *Juniperus osteosperma* reaches only 12–20 feet (3.5–6 m) in height, though stout-trunked. Then there are many trees of highly infertile or waterlogged sites, such as the coastal dunes and bogs of eastern USA. Yet another category is the understory trees in taller forests or woodlands, found in any wooded region. The yews (*Taxus*) are north-temperate examples. It can be supposed that many of the more adverse environments impose limits on tree size; conversely, many species have evolved a size to match these limits.

The old and the new

All present-day trees are both old and new in the evolutionary sense. They are all descendants of the earliest land plants, yet are all modern representatives of their botanical groupings. Some are seen as "old" because they belong to groups that were more abundant in past geological eras and are now largely or totally extinct, such as the cycads, *Ginkgo* and some conifers. But none of these is identical to its Jurassic or Cretaceous ancestors of over 100 million years ago. The flowering plants (angiosperms or magnoliophytes) are more ancient than once thought, though their chain of descent prior to the Cretaceous remains a mystery. It is still uncertain which fossil plant groups from the Triassic and Jurassic periods were their most likely ancestors, or whether any fossils yet discovered represent those ancestors. What is known with reasonable confidence is that by the mid-Cretaceous (around 100 million years ago) a large proportion of present-day flowering-plant families existed. Some families appeared in the early Cretaceous, before the major split of the flowering plants into the dicots and monocots. These families are recognized as the "primitive" or basal angiosperms. Right from the beginning it appears that these families included a high proportion of trees, although it was probably not until the end of the Cretaceous or later that flowering trees began to dominate some of the world's forests. The change to drier climates in the mid-Tertiary, around 30 million years ago, stimulated major bursts of flowering-plant diversification in groups such as the legumes, daisies and grasses.

↓ **This cross-section through petrified wood** of Jurassic age from Utah comes from the extinct tree genus *Hermanophyton*, but it is not known what its leaves and reproductive organs looked like. One small group of seed ferns probably gave rise to flowering plants.

EARLY PLANT LIFE

Tall treelike plants first dominated Earth's vegetation during the late Devonian–early Carboniferous periods, more than 350 million years ago. They consisted largely of plants known as pteridophytes or "fern allies," that reproduced by spores. These plants are now considered as an evolutionary "grade," not a single natural group. Prominent among them were the lycophytes or "giant club mosses" and sphenopsids or "giant horsetails," some up to 100 feet (30 m) tall. Most families had become extinct by the Jurassic and only small remnants survive to the present day. Seed-bearing plants were evolving at almost the same time. The "seed ferns," with seeds borne on edges of fernlike fronds, appeared in the late Devonian and were diverse in the Carboniferous and Permian. These were the first gymnosperms (naked-seeded plants), a class which, like the pteridophytes, is now regarded as only an evolutionary grade. Survivors are the conifers, cycads, gnetophytes and *Ginkgo*.

↑ **Prehistoric forests** were dominated by pteridophytes ("fern allies") and horsetails.

→ **This classic evolutionary chart** illustrates the history of vascular land plants from the Silurian period to the present-day. The differing widths of the "branches" are a guide to the variations in abundance and diversity of each group, as they have changed through time.

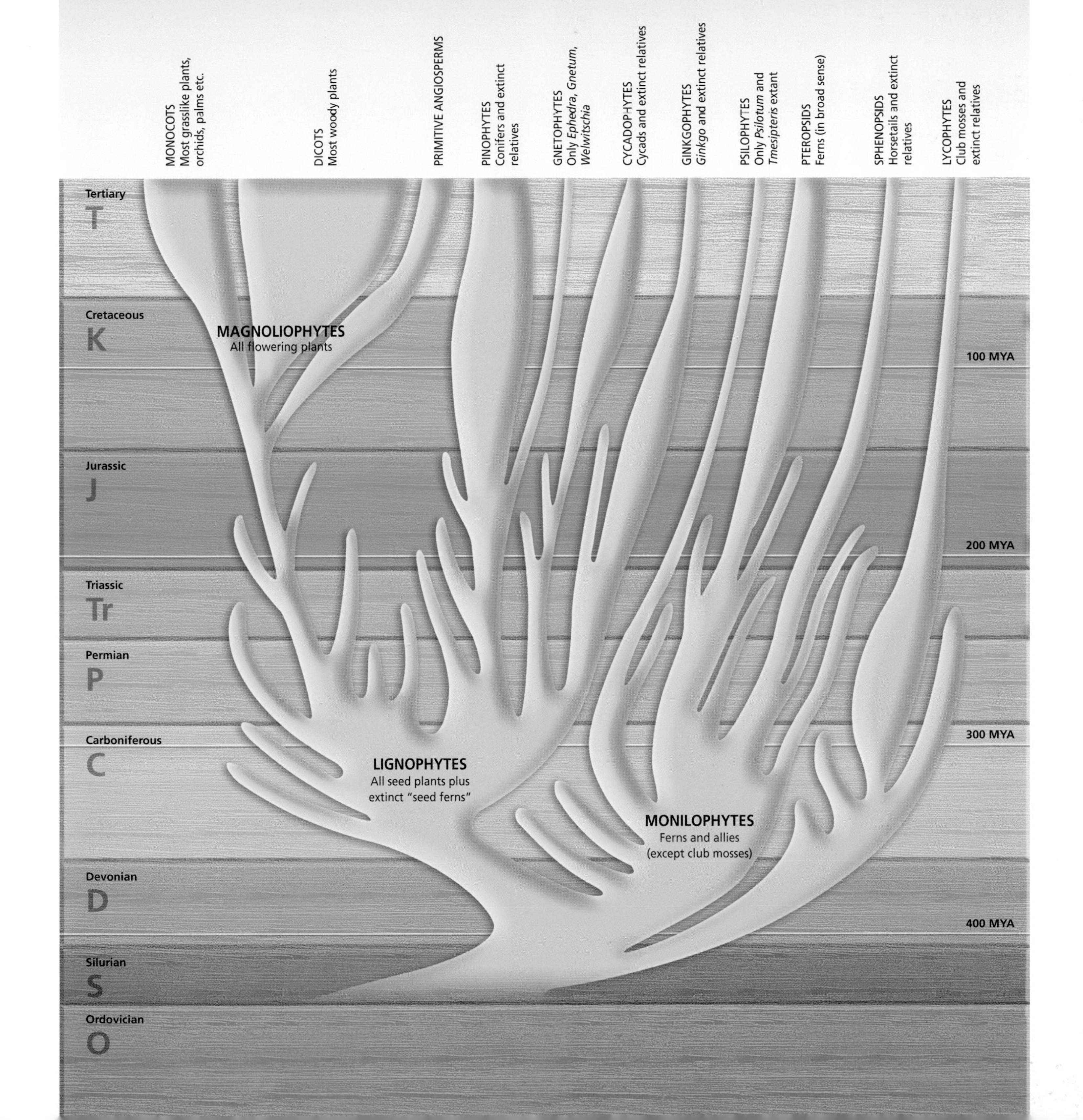

MONOCOTS
Most grasslike plants, orchids, palms etc.
DICOTS
Most woody plants
PRIMITIVE ANGIOSPERMS
PINOPHYTES
Conifers and extinct relatives
GNETOPHYTES
Only Ephedra, Gnetum, Welwitschia
CYCADOPHYTES
Cycads and extinct relatives
GINKGOPHYTES
Ginkgo and extinct relatives
PSILOPHYTES
Only Psilotum and Tmesipteris extant
PTEROPSIDS
Ferns (in broad sense)
SPHENOPSIDS
Horsetails and extinct relatives
LYCOPHYTES
Club mosses and extinct relatives
Tertiary
T
Cretaceous
K
Jurassic
J
Triassic
Tr
Permian
P
Carboniferous
C
Devonian
D
Silurian
S
Ordovician
O
100 MYA
200 MYA
300 MYA
400 MYA
MAGNOLIOPHYTES
All flowering plants
LIGNOPHYTES
All seed plants plus extinct "seed ferns"
MONILOPHYTES
Ferns and allies
(except club mosses)

The art of naming trees

Like all organisms, trees are given names depending on how they are classified; and they are classified depending on the variation that can be demonstrated to exist among them. This last part is the realm of science—discovering what is really "out there" and what are the degrees of resemblance or difference among all the individual trees. Once these patterns of variation are established, by field and laboratory work and statistical analysis, biologists draw boundaries around groupings of individuals that most resemble one another. These initial groupings are called species. Wider boundaries drawn around groups of species are known as genera (plural of genus); and so on, through higher levels of classification such as family and subdivision. Naming is the final, though scientifically least important, part of taxonomy, the discipline that encompasses these activities. Taxonomists give organisms names of their own choosing and the first name published usually has priority. In plant taxonomy, names are based on Latin or latinized Greek, an ancient practice in European botany though codified by the Swedish taxonomist Linnaeus in the mid-eighteenth century. Linnaeus gave each species a binomial, consisting of the name of its genus followed by a "specific epithet." Thus in the genus *Pinus* we have the species *Pinus pinaster*.

↗ **Conifers, together with cycads and the Ginkgo** are all gymnosperms. These are plants that reproduce by means of a seed that is exposed, often on the scales of a cone or similar structure.

→ **Angiosperms (or flowering plants)** are plants that reproduce by means of a seed that is enclosed in an ovary.

← **Linnaeus (Carl von Linné)** laid the foundations of scientific classification when he classified all plant species then known in his *Species Plantarum* (1753). This volume is now accepted as the starting point for priority of botanical species names.

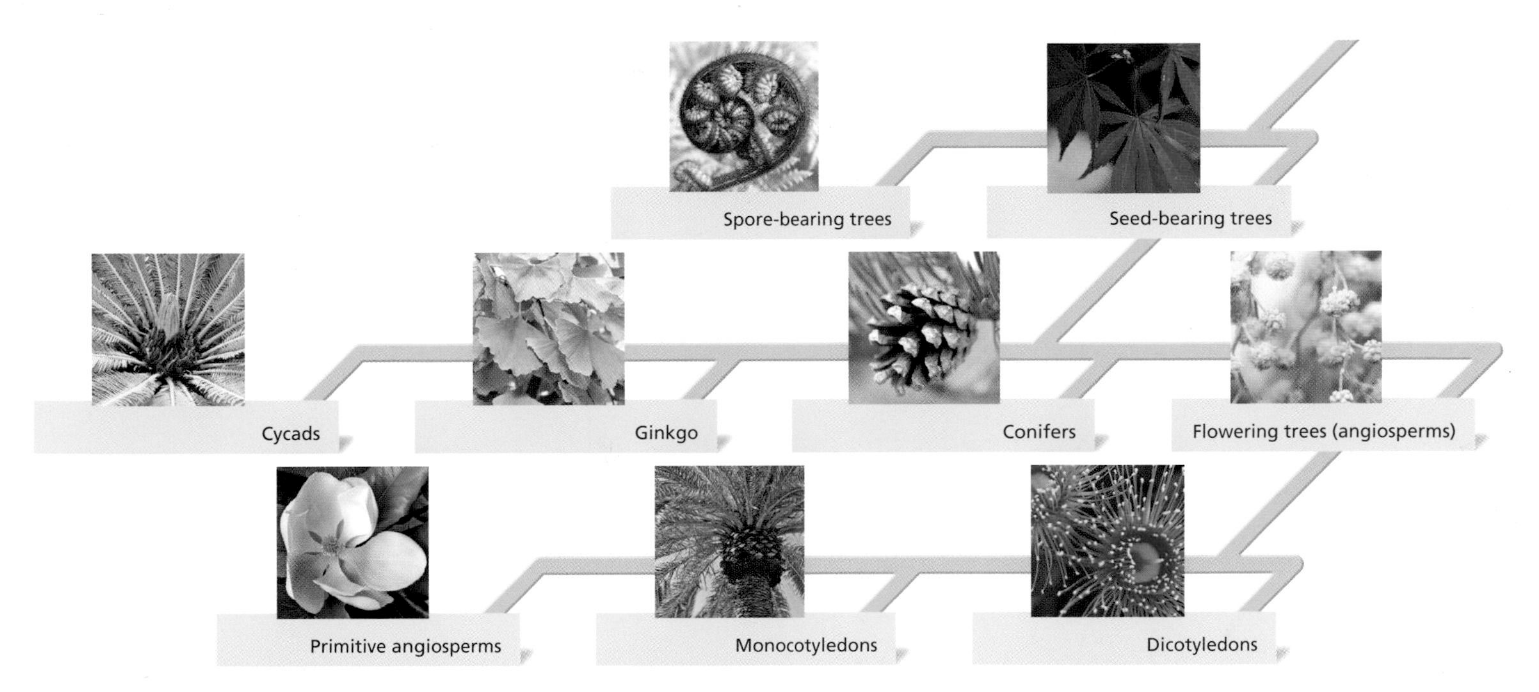

↑ **The major groups and subgroups of trees.** Note that trees are very unevenly distributed among the groups, from Ginkgo with its sole species *Ginkgo biloba*, to the astonishing variety of the dicotyledons. There are well over 300,000 species of flowering plants (angiosperms), while there are only 600 or so species of conifers. However, conifers cover the Earth in numbers that are vastly out of proportion to their diversity.

BOTANICAL HIERARCHY

Botanical classifications are hierarchical. A species belongs to only one genus, a genus to only one family, a family to only one order, and so on through all the levels of classification to the plant kingdom. Going from the top downward, an order may include many families, a family many genera, and so on. Above the level of genus, rules of nomenclature govern the form of scientific names, such that their endings indicate their position in the hierarchy. All family names have the ending -aceae, subfamily names -oideae and order names -ales. A family name incorporates the name of one of its component genera, or at least its "stem." Thus the family Pinaceae includes the genus *Pinus*. The name is carried on upward through the order Pinales and the division Pinophyta, and downward to the subfamily Pinoideae.

↓ **Hierarchy of classification.** The chart below demonstrates the traditional classification of one Rhododendron, working downward in descending hierarchy from Plantae (the whole of the plant kingdom) to the specific cultivar.

KINGDOM **Plantae** all plants
DIVISION **Magnoliophyta** all flowering plants
CLASS **Magnoliopsida** dicotyledons
SUBCLASS **Asteridae** "asterids," a large group of families in APG
ORDER **Ericales** smaller group of families
FAMILY **Ericaceae** heath family
SUBFAMILY **Rhododendroideae** Rhododendron subfamily
TRIBE **Rhododendreae** Rhododendron tribe
GENUS ***Rhododendron*** Rhododendron and Azalea genus
SUBGENUS ***Hymenanthes*** most tall evergreen rhododendrons
SECTION ***Ponticum*** only section of this subgenus
SPECIES ***Rhododendron arboreum*** Indian Rhododendron
SUBSPECIES ***Rhododendron arboreum* subsp. *cinnamoneum*** Nepalese race of Indian Rhododendron
CULTIVAR ***Rhododendron arboreum* subsp. *cinnamoneum*** 'Blushing Beauty' selected clone of Nepalese race with white flowers flushed pink

Climate and trees

Climate is by far the most important factor in determining what type of vegetation grows in any given location in the world. Climate is not the same as weather, but is the sum total of all the weathers through each year and over many years. Climate can be summarized in statistics such as annual rainfall, average and extreme temperature, humidity and windspeed for the different seasons, and incidence of frosts and snow. In the case of rainfall, its distribution through the year is almost as important as its total amount: regions of winter-dominant rainfall support different kinds of forest from those where rainfall is summer-dominant. Climate is determined by factors such as distance from the equator, height above sea level, and proximity to oceans or seas with varying surface temperatures. In the northern hemisphere, where the two major continents almost encircle the Earth in their arctic regions, the southerly outflow of very cold air masses that occurs from time to time in winter, affects climates at much lower latitudes, such as in central Europe, eastern China and central USA, producing much lower winter minimums than at equivalent latitudes in the southern hemisphere.

MICROCLIMATE

Within a continent or large island, climates vary mostly from one broad region to another, though a narrow coastal band may be regarded as a different climate zone from its hinterland due to the modifying effect of the sea; likewise a plateau or mountain range, because of cooling at higher altitudes and interception of rain-bearing clouds. But there can be variations over much shorter distances, and these are called microclimates. In temperate latitudes, the northern and southern slopes of a hill, for example, will have warmer and cooler microclimates due to the sun's angle; and a valley bottom will have a more frosty microclimate due to cold air drainage at night. Forests produce their own microclimates beneath their canopies.

← **Lowland tropical rain forest,** such as this example on a tributary of the Amazon River, develops only with sustained warm (but not extreme) temperatures and year-round rain with accompanying high humidity. Most occurs close to the equator.

⇇ **On the rocky Utah plateau** in Zion National Park, pine woodland is sparse and stunted due to lack of soil, but climate and microclimate have played a part. At up to 10,000 feet (3000 m) altitude, the plateau has very cold winters and receives little rainfall.

← **This tropical coral islet** supports a dense stand of small trees in its center. Its climate is constantly warm, the temperature varying within a narrow range; but it is vulnerable to occasional cyclonic winds.

⇇ **Occurring at a similar latitude** to the French Riviera, these forested ranges in Jilin Province, northeastern China, experience lower winter temperatures due to arctic air outflows, and the forests contain some boreal tree species, such as larches.

Climate and trees continued

Every one of the vast number of tree species in the world is adapted to certain climatic conditions, regarding both the limits beyond which it cannot survive, and the climate that is optimum for its growth. Climatic tolerances of a tree are essentially physiological, dependent on the chemistry and physics of the living cells and the sap fluids exchanged between them, although anatomical structures are also important. These tolerances have evolved over many millions of years and are not easily modified, for example, by domestication and breeding. Thus, no orange tree can survive winter outdoors in New York State despite more than a century of effort by plant breeders. Foresters may seek out what they call "provenances" of a tree species, originating from the coldest parts of its natural range, but these will only survive temperatures a very few degrees lower than what any other trees of that species can survive. In the other direction, trees from cool climates do not thrive in warm ones, in some cases being killed outright by the first blast of summer heat—though more often they languish and fail to flower or fruit, due to absence of the seasonal changes needed to trigger leaf burst, flower initiation or fruit set.

GEOGRAPHICAL RANGE

The geographical range of a tree species in the wild is not determined only by its climatic tolerance. Many factors may combine to keep it to a smaller range, the most obvious being barriers such as oceans or mountain ranges. Just as important is competition with other tree species, and present day distributions are the result of a complex history of events over many thousands or millions of years, as species jostle with one another for room, climates shift, and changes in the Earth's surface, such as uplift, volcanoes and glaciers, create both barriers and new lands to be colonized. The climatic warming we are experiencing at present will eventually shift the ranges of many trees and other plants.

→ **Deciduousness in a tree species** is generally an indication that it is adapted to a strongly seasonal climate—either with severe frosty winters and warm summers, or with a tropical dry season. In both cases, leaves may change color before falling. These American Aspens (*Populus tremuloides*) are among the most frost hardy deciduous trees.

↙ **Trees that live in areas with very cold winters** may become adapted to coping with ice storms. While some species may be killed, others will tolerate ice forming around the stems.

↓ **There are a number of ways in which trees** become adapted to arid climates, and one of them is succulence. These Quiver Trees (*Aloe dichotoma*) in Namibia store water in both the succulent stems and leaves, allowing the tree to survive long periods without rainfall.

Form and function

Previous page The Chinese or Lace-bark Elm (*Ulmus parvifolia*) is a semi-deciduous tree native to China, Japan, North Korea and Vietnam.

Form and function

Like other plants, trees basically consist of shoots, stems, leaves and roots. Many trees have tall trunks, allowing them to compete more effectively for sunlight so the leaves can photosynthesize and produce food for the plant. Some trees have developed specific adaptations to deal with the particular vagaries of their climate.

Tree structure	34
Trunk	36
Branches and twigs	38
Wood	40
Bark	44
Roots	46
The soil	48
Leaves	50
Fertilization and reproduction	56
Flowers	58
Cones	64
Fruits and nuts	66
Seeds	68
Tree shapes	70
Growth stages	72
Tree growth	74
Attack and defense	76
Survival of the fittest	80
Wind and weather	82
Fire and flood	84
Tree pioneers	86

Tree structure

As with all plants, the specific structure of a tree is designed to maximize its growth and chances of successful reproduction. A tree is composed of roots; a trunk which is generally protected by layers of bark; branches and twigs; and leaves. Unlike humans and animals, trees and other plants have a series of localized functions rather than a central control system (like animal's nervous system). Since trees are restricted in their movement, their size and shape have evolved to favor growth and reproduction in various environmental niches. A tree's tall trunk and spreading branches expose its leaves to the sun and air, and its flowers to pollination—either by wind, insects, birds or animals. Beneath the ground, the tree's root system forages in the soil for moisture and nutrients, its branching shape often mirroring that of the true branches aboveground. The leaves produce food for the tree through the process of photosynthesis. Trees are generally larger in size, in both height and spread, than other types of plants, such as shrubs, vines or perennials, since they have a large woody trunk, which divides into, or gives rise to, branches. The growth habit of a typical shrub is to have many stems arising from the base but trees generally have a single trunk. However, the distinction between shrubs and trees can blur when trees are multi-stemmed or produce suckers or stems from multi-growth points at root level.

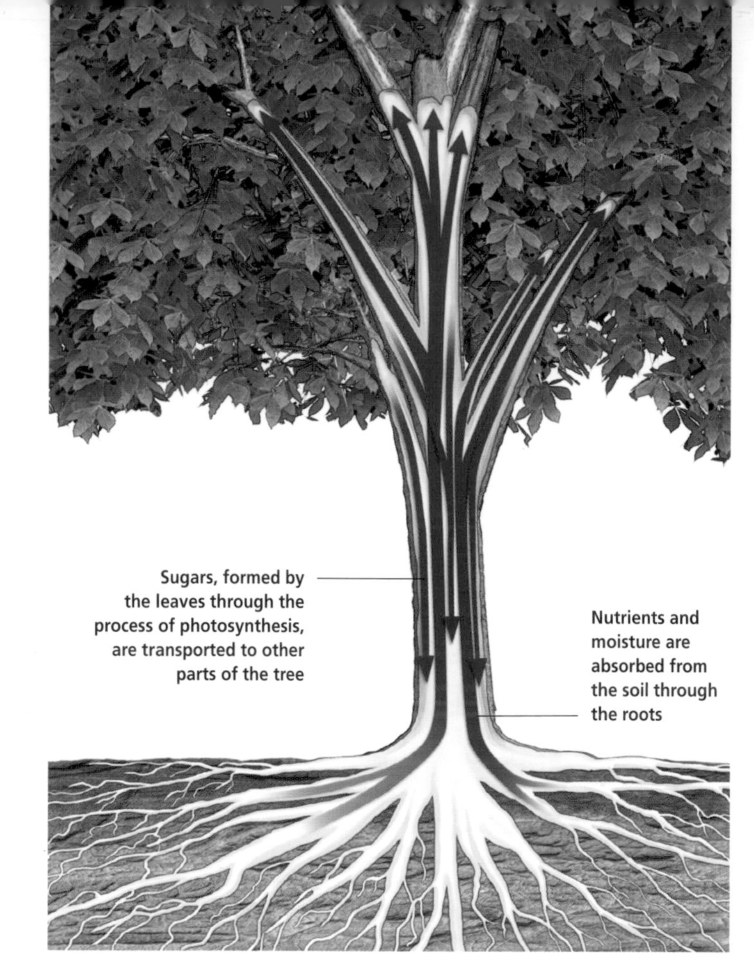

↑ **Transpiration is the plant equivalent of respiration** in animals. Moisture is taken up from the soil via the root system and carried to all parts of the tree through specialized cells called xylem.

↗ **A typical tree** is of large size and has a single tall or stout trunk covered in bark, as well as spreading branches emanating from the trunk.

← **Tree ferns are early treelike plants** that appeared on earth some 300 to 350 million years ago. As tree ferns grow, their fronds are discarded leaving distinctive markings on the trunk.

VARIATIONS FROM THE NORM

Although a typical tree consists of a single main trunk clad with bark, from which appear branches that spread out to form a canopy of twigs and leaves, some trees have a different growth pattern yet are still described as trees. Trees growing in arid deserts, in fire zones, or close to the coast may be small or may develop a multi-stemmed growth habit in response to the environment. Although having more than one trunk, they take on treelike proportions as they grow, and also have bark and a woody, self-supporting trunk. Most palms have a single, straight trunk and a crown of foliage. However, they differ from other trees since they have no bark. Some ferns reach 33 feet (10 m) tall and have fibrous trunks, but they also lack true bark.

→ **Plant cells** have a cell wall (shown as yellow) and chloroplasts (green). The chloroplasts contain a green pigment, chlorophyll, that carries out photosynthesis to provide the energy for plant growth. Plant cells have a large central space called a vacuole that is filled with liquids and gases. The nucleus (purple) contains nucleolus and is common to both plant and animal cells.

← **Apart from the common element** of a nucleus (purple), animal cells have a different shape and structure from the typical plant cell. The nucleus in an animal cell contains the building blocks of life, DNA, that control protein synthesis. Mitochondria (the cylinders such as the one seen bottom center) generate energy for the animal cell to function.

Trunk

The distinguishing feature of most trees is the trunk, which allows the tree to grow upright so the foliage can reach the sunlight. The higher up the branches and leaves, the less competition there is from surrounding plants. The trunk is normally single and self-supporting, and can range from a few inches to over a hundred feet in length. This length difference is determined by the tree's habitat as much as by genetics. The same species, when growing in an exposed and harsh position, can have a short trunk while, if growing in a more favorable habitat, it can have a much longer trunk. Some specialized tree species have a trunk that is mostly underground, with trunklike branches growing almost vertically. The majority of trees have a trunk that is without branches for varying distances from ground level. Many trees have branches for the whole length of the trunk for their lifetime, while others shed their branches as they grow taller, ending up with a canopy of branches high above the ground. Depending upon the climate, the trunk, with its outer covering of bark, is often home to many other plants, such as epiphytes, and to many species of animal, mostly insects and arachnids, that live on or under the bark.

→ **The tropical Batai (*Albizia falcataria*)** is one of the fastest-growing trees in the world, increasing in height by up to 25 feet (7 m) per year, eventually attaining 110 feet (30 m) in 25 years. It is a short-lived, deciduous, pioneer species, colonizing gaps in forests after damage by hurricanes.

BAOBAB TREE
The baobab tree of the genus *Adansonia* is native to Madagascar, mainland Africa and Australia, where it may also be called the boab or bottle tree. The massive swollen trunk stores water that allows the tree to survive very dry periods, and it is possible for it to contain up to 30,000 gallons (120,000 liters) at one time. The enormous trunks of the baobab are sometimes hollowed out and have been used by humans for anything from water reserves and shelters to a small prison. The baobab trunks are also of interest to animals, and elephants use their massive tusks to dig into the tree to seek the moisture and pulp inside.

↑ **The Monkey Puzzle (*Araucaria araucana*)** grows to 130 feet (40 m), with a trunk diameter of over 6 feet (2 m). Since the tree grows in regions of heavy snowfalls, the older branches are broken off by the weight of the snow, culminating in an umbrella of branches, with the branch scars often prominent.

↗ **The American Aspen (*Populus tremuloides*)** is a deciduous tree that grows to 50 feet (16 m). It has a slender trunk with yellow to gray bark.

→ **Palms do not have a woody trunk** like most other trees. Rather they are composed of soft tissue in which are embedded many strands of vascular tissue, through which water and nutrients pass. (This unique structure of a palm trunk inspired the engineering design of cable-reinforced concrete.) Palm trunks are extremely flexible, being able to withstand hurricane-force winds. The patterns on the trunk pictured are the scars left when the fronds fall. These scars are often unique to particular species.

→→ **All species of baobab** have a swollen trunk, often up to 30 feet (9 m) in diameter and up to 50 feet (15 m) in height. When the trees are leafless, usually in the dry season, they give the appearance of an "upside down" tree. The wood is quite soft and contains vast amounts of moisture.

Branches and twigs

The secondary limbs, or branches, are the aerial and lateral extensions that grow from the trunk. They further enable the tree to reach out to the sunlight, and also help in the transportation of water and nutrients. The branches form the canopy of the tree and produce smaller growths called twigs, from which flowers and leaves grow. Branches always remain at the same distance from the ground, simply becoming thicker each year and forming thick wood as support. The shape of trees formed by the arrangement of branches varies considerably and can reflect the climate in which the tree grows. For example, in colder parts of the world where snow falls, evergreen trees such as European Larch (*Larix decidua*) or Norway Spruce (*Picea abies*) are generally tall and straight, their branches arising from the main trunk and angled downward to enable snow to be shed to avoid it weighing down and breaking off the branches.

↓ **In spring the new growth on deciduous trees** bursts into life from the buds that have remained dormant through the winter cold.

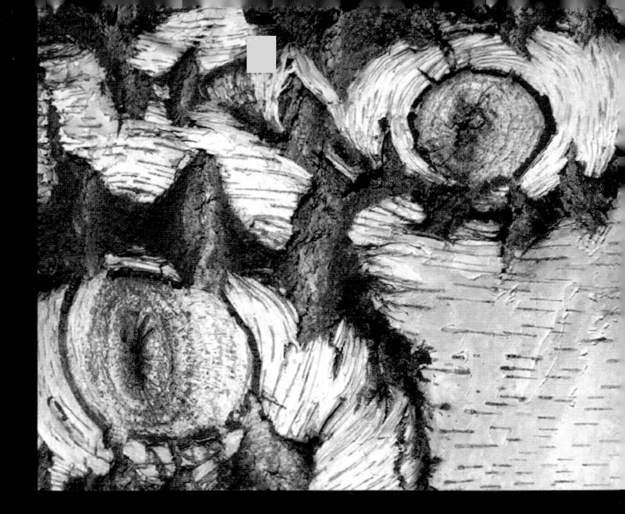

↑ **Scarring occurs on the trunk** where a branch or twig has been discarded, forming distinctive patterning known as "eyes." Wounds naturally seal, thereby preventing disease from entering the tree.

← **Branching patterns vary from tree to tree.** On this fir tree the branches arise regularly around the trunk, forming a conical shape, which is typical of many conifers.

↞ **The Joshua Tree** (*Yucca brevifolia*) has distinctive contorted branches topped with tufts of sharp leaves.

← **Branches often grow to form a tree** with a dense, spreading canopy that shades the trunk and soil from the sun's rays.

↞ **This Monkey Puzzle Tree** (*Araucaria araucana*) has formed a broad crown at the top of its trunk. The tough, sharply pointed leaves overlap one another along the branches for protection and flexiblity.

BRANCH MODIFICATIONS

Branches and twigs normally carry leaves. However, some trees have branches that are modified with defensive thorns that serve as protection against predators. Some trees may dispense with normal leaves altogether and form leaflike cladodes or phyllodes. These structures are adaptations that allow the tree to survive hot, dry conditions by reducing the rate of transpiration (the tree's natural lose of water). These modified branches can be mistaken for true leaves and characterize many trees growing in dry or barren conditions, such as cacti as well as acacia and mimosa trees.

Wood

As well as giving a tree the support and rigidity it needs to grow and develop, its trunk contains tissue through which water and nutrients are distributed and stored as starch. The bark is the outermost layer of the tree. Underneath this outer, protective layer is the inner bark, cambium, sapwood and, in the center of the trunk, the heartwood. It is in the area of the cambium that new cells are formed, leading to the growth of the girth of the trunk. This growth is termed exogenous, meaning outward growing. The cells produced in the cambium add to the amount of sapwood in the tree. As more cells are added to the outer edge of the sapwood layer, the tree trunk grows larger. As the area of sapwood expands, the older sapwood toward the center of the trunk becomes less active and is referred to as heartwood. This type of growth is very different from the way palms grow. Their growth is termed endogenous, meaning growing from within. As a palm grows, the diameter of its trunk changes, reflecting the amount of moisture and nourishment available. In poor years, the diameter of a palm's trunk is smaller but it expands in good years.

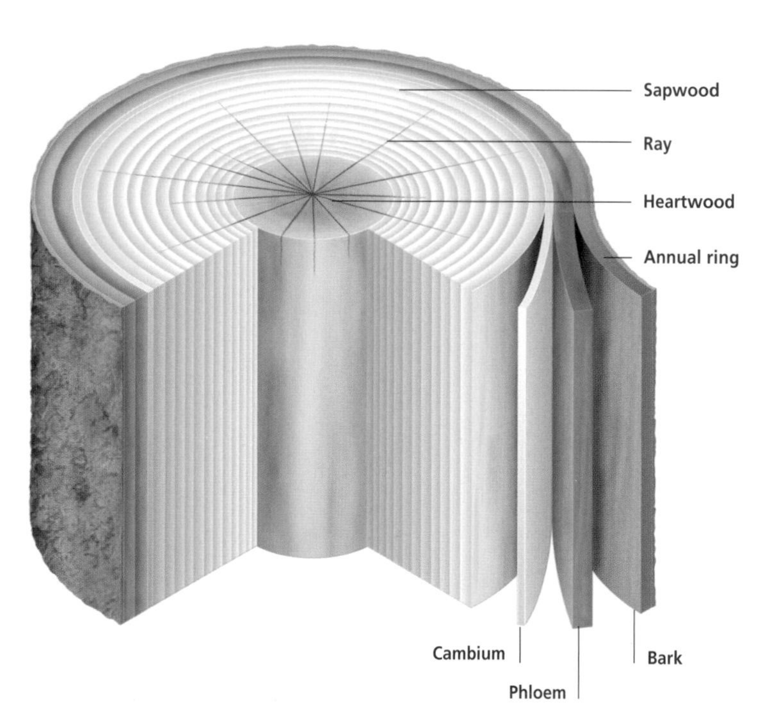

↑ **This cross-section of a trunk** illustrates the different types of wood: bark, phloem, cambium, sapwood and heartwood. Each of the ring patterns through the trunk denotes a year of growth.

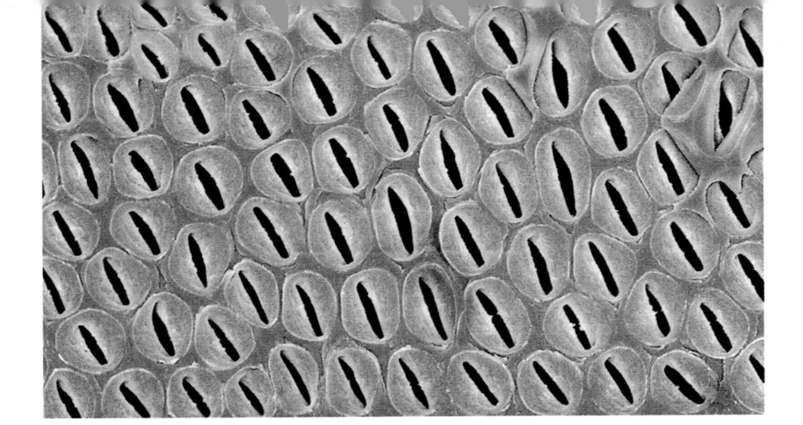

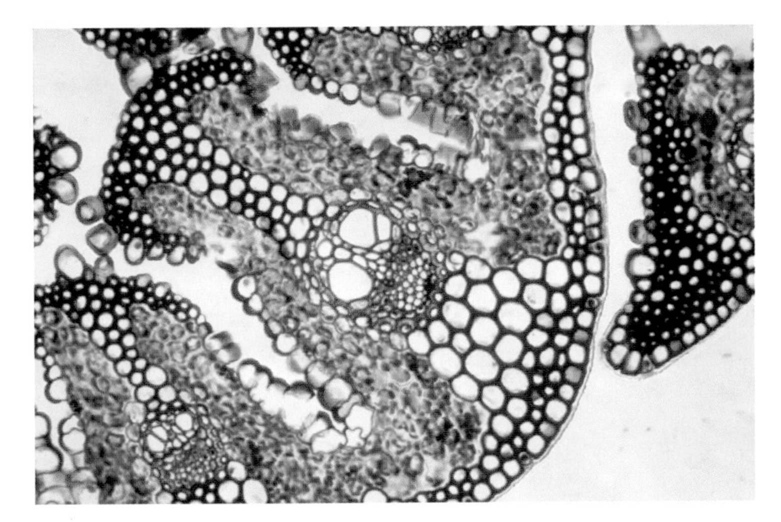

↟ **Seen through a colored scanning electron microscope the xylem,** the plant cells that transport water and minerals from the roots, are revealed as bundles of tracheid cells each with a wall thickened with lignin (seen as green).

↑ **Under a microscope, sclerenchyma cells** are revealed with their thick, lignified walls. These cells strengthen the tree, sometimes giving rise to fiber that's strong enough to make into ropes and fabrics such as sisal.

← **This cut lumber reveals the structure of the trunk.** Beneath the outer layer of bark the pale sapwood and darker heartwood are evident.

XYLEM AND PHLOEM

The sapwood is the powerhouse of the tree. It is through this actively growing tissue that moisture and minerals are carried. The cells involved in the transportation are known as the xylem and phloem. The xylem conducts moisture and dissolved minerals from the roots upward and throughout the plant. The cell walls of the xylem are strengthened with a fibrous material called lignin. The phloem is the area of the sapwood responsible for the transportation of sugars. These sugars are manufactured by the leaves in a process called photosynthesis and then carried down the plant through the cells of the phloem. Sugars are transported the full length of the tree from its leaves to its roots. Cells known as rays, found within the phloem, also carry sugars inward into the sapwood where they are converted to starch and stored.

Wood continued

The hard, darkened wood that lies at the center of trunks and branches is termed the tree's heartwood. It can be clearly seen if the trunk or branch is cut. This dark area was once xylem and phloem, involved in the transportation of nourishment to all parts of the tree. As sapwood cells break down with age and use, and the tree forms new sapwood, the heartwood tissue takes on a new role, giving the tree strength and rigidity. The darker coloration seen in heartwood, compared to the surrounding sapwood, is due to its impregnation with oils, gums and resins. These substances help to protect the heartwood from insect attack or fungal decay. As a tree ages, its heartwood is more likely to be damaged. It is possible for the heartwood to be destroyed by fire, rot or insect attack. Despite this damage, the tree is able to continue to survive and grow. However, if the outer layers, particularly the cambium and sapwood, are damaged, for example by ring barking, the tree dies. Such old hollow trees, often referred to as pipes, are found in old growth forests where they provide homes for birds and animals that nest or shelter in trees. In Australia many old eucalypts form pipelike structures after fire has destroyed their heartwood.

→ **A scanning electron microscope** reveals the cells in the center of a trunk of English Oak (*Quercus robur*). This cell, once forming part of the xylem and carrying water from the roots to the rest of the tree, is blocked with a papery growth called tyloses. Surrounding the cell is tissue called parenchyma.

↓ **Growth variations** occurring through the life of a tree are apparent when its trunk or branches are cut. The two distinct growth points seen in this piece of wood were caused by damage to the original growing tip. When the timber is to be used as a building material, growth irregularities, such as frequent branching, splits, multiple growing points and insect or disease damage, can reduce the amount of lumber that can be cut from a log. However, heavily figured timber can be attractive.

HARD AND SOFT TIMBERS

One of the most beautiful of all heartwoods is found in the tropical rain forest tree, Ebony (*Diospyros ebenum*), which is black, exceptionally hard and has a fine grain. This wood is used to make the black keys of a piano, to inlay fine furniture and for decorative carvings, such as chess pieces or jewelry. The timber from rain forest trees like ebony, evergreen temperate forest trees such as eucalpyts, and from deciduous trees is collectively referred to as hardwood. The term softwood is reserved for timber cut from coniferous trees, such as pines and spruce. The terms softwood and hardwood refer to the water-conducting cells, and not to the actual softness or hardness of the wood.

↓ **Branches arise from the trunk** creating a change in the regular growth pattern which, in cut lumber, is called a knot. Here the branch has been reduced to a stub by pruning. The tree's bark has grown around the dead branch, sealing the wound to restrict entry to its sapwood of unwanted insects or diseases.

↓ **Section of softwood.** In softwoods, the water-conducting cells are known as trachaeids and are tapered in shape.

↓ **Section of hardwood.** In hardwoods the water-conducting cells are tubular-shaped and are known as xylem vessels.

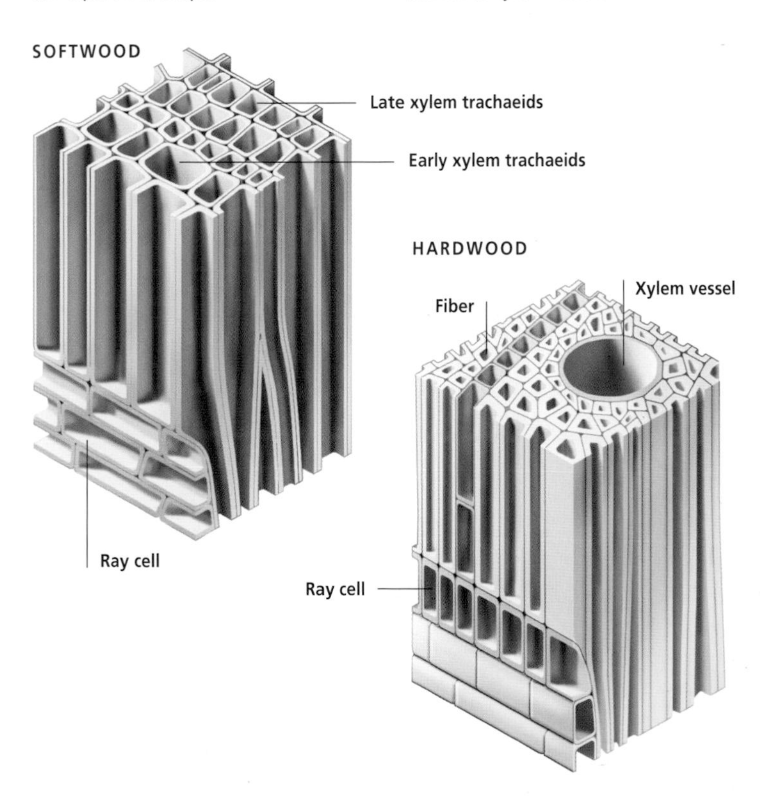

Bark

Bark is one of the most obvious features of a tree and is the outer layer that is exposed to the elements and is considered to be dead. The bark protects a tree against climate extremes and damage by animals or pathogens. For some tree species bark is a fire shield. Thick and dense around the trunk it may char during a fire, but the living wood beneath remains alive and healthy. Some trees that rely on the protection of their bark to survive fire include many eucalypts (*Eucalyptus sp.*)and the Californian Redwood (*Sequoia sempervirens*). To understand the vital protective role of bark, it is necessary to look beneath the surface. Under the bark, known as the outer bark, lies the vital living and growing tissue, known as the inner bark. In this part of the tree is a system of conductive cells called the phloem, which contains a stream of nutrients coursing through the plant from the leaves to roots. In some trees, bark is shed in plates to make way for regrowth. This process may occur annually, often during a period of active tree growth. However, not all trees lose their bark, some barks stretch to accommodate tree growth, becoming extremely smooth where the bark is stretched around the trunk; or fissured, furrowed or fibrous and soft as the bark ages and expands.

→ **The peeling bark on a Manzanita tree** (*Arctostaphylos manzanita*) transforms this tree into a living sculpture. Many trees shed their bark annually, revealing fresh, boldly colored bark beneath. This species of manzanita is native to California, where it is grown as a garden and street tree. Other trees noted for peeling or decorative bark include maples, including the Snake-bark Maple (*Acer grosseri* var. *hersii*), which has green bark striped with white; birch trees such as the Black Birch (*Betula nigra*); and the Paperbark (*Melaleuca quinquinervia*), which is native to Australia.

OTHER BENEFITS OF BARK

As well as serving to protect and insulate a tree, some bark is highly ornamental. Cherry trees such as the Manchurian Cherry (*Prunus maackii*) have smooth, red bark that is often attractively marked by leaf scars. In many areas of the world bark is an important by-product of the logging industry. Once it is stripped from the logs (near the point of harvest or at the sawmill), it is sold on to composting specialists to provide a soil alternative, known as potting mix or compost. Bark is also prepared as a mulch to spread around garden plants for extra protection against drought or extremes of temperature. The bark of the Cork Oak (*Quercus suber*) is harvested then manufactured into corks to stopper wine bottles or as a soundproof substance used for flooring or insulation. Some popular spices, including cinnamon (from *Cinnamomum verum*) are obtained from bark. In a two-year process, young cinnamon trees are coppiced (cut to ground level) so the bark is harvested from the new stems.

CROSS-SECTION OF BARK SHOWING DEVELOPMENT OF CORK CAMBIUM

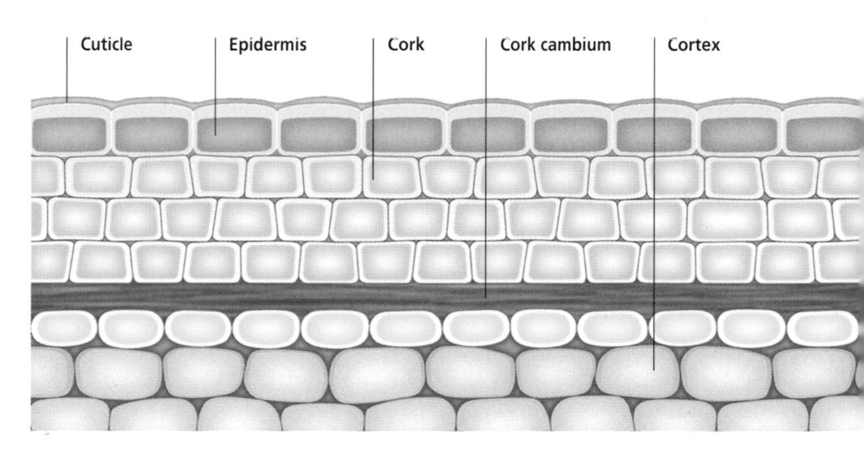

↓ **The Tibetan Cherry** (*Prunus serrula*), has a copper-red trunk that is particularly striking in late fall and winter when the tree is bare and the trunk exposed.

↓ **The bark of many eucalypts** is shed to reveal colorful new bark underneath. The long strands of discarded bark build up around the trees.

↓ **The deeply fissured bark** of this Shortleaf Pine (*Pinus echinata*) forms platelets around the trunk as the tree ages.

Roots

The root system of a tree has two major roles. The roots anchor the tree in the ground and also forage in the soil for minerals and moisture to supply the tree as it grows. Without moisture supplied via the root system a tree wilts and eventually dies. The roots also become storage depots for moisture and nutrients, stored up in times of plenty to supply the tree in poor seasons. Most trees have large, spreading root systems that provide anchorage to the tall, leafy, above-ground structure. This type of root system is known as a fibrous root system. Most roots grow downward under the influence of gravity and away from the light. However, many trees roots live in a relatively shallow layer of soil. At the tip of all growing roots is a mass of cells, termed the root cap, which protects the root as it grows and travels through the soil. These cells are actively growing all the time and, as cells wear away, they are quickly replaced. The root is forced through the soil by lengthening cells just behind the root tip. These cells lengthen rapidly, giving a growing root considerable force to push through the soil. Behind the growing tip, the root hairs sprout. These minute growths absorb soil moisture. They are constantly replaced and regrown as the root extends. The internal structure of the root, with conducting cells, reflects the structure of the aboveground tree.

→ **Buttress roots** provide extra aboveground support for large trees and arise from the base of the trunk. They are features of tropical trees like this *Ficus macrophylla*, which develop large, spreading canopies. The Kapok Tree (*Ceiba pentandra*) also forms substantial buttresses.

ABOVEGROUND ROOTS

Many trees have secondary roots aboveground. Some of these are aerial roots found, for example, on figs. Aerial roots arise from branches, take in moisture from the air and, when they reach the ground, provide the branch with additional support. In very wet soils or when trees such as mangroves are growing in water, the root system has developed ways to cope with inundation. Such features include peg or stilt roots and pneumatophores (kneelike aerial roots that rise above water), which enable the tree to continue to obtain oxygen. Pneumatophores are characteristic of trees growing in swampy conditions such as the Swamp Cypress (*Taxodium distichum*).

← **Aerial roots sprout high above the ground** from the stem on this Banyan Tree (*Ficus benghalensis*) where they grow downward to the ground forming trunklike supports for the heavy, spreading branches.

→ **The Swamp Cypress,** which grows in water in parts of southern USA, has structures called pneumatophores which enable the roots to take in oxygen. Without these structures the submerged roots would die.

CROSS-SECTION OF A ROOT

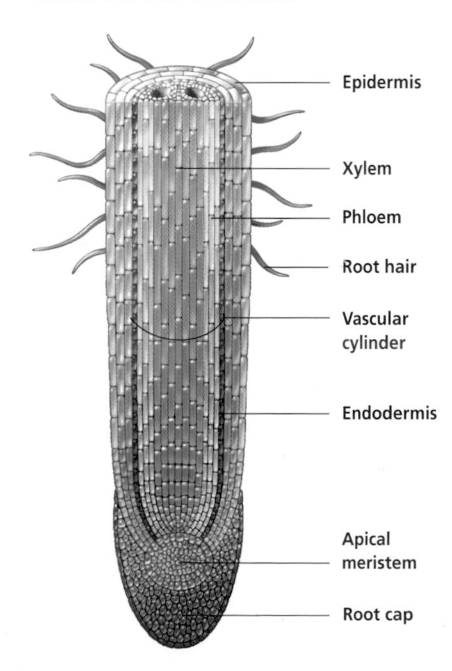

The soil

Soil is the topmost layer of Earth's crust and is the material in which trees grow and spread their roots. Soil is made up of particles of earth, such as sand, clay, chalk or gravel, together with voids that contain air and water. A balance between air and water in the soil provides an ideal growing medium for tree roots. Soil also contains nutrients, such as nitrogen, along with minerals, such as phosphorus and magnesium, which are required in varying amounts for plant growth and are absorbed by the tree's roots. Soils change in composition and appearance with depth, creating what is known as a soil profile. A typical soil profile where trees grow, such as in a forest, has a top layer of humus that is formed by fallen leaves and debris that have gradually decomposed. Below the humus is the topsoil, which can range in depth from a few inches to several feet, and this is where most tree roots are concentrated. Under the topsoil is the subsoil, which is generally lacking in humus and therefore has poorer nutritional value for plants. Below the soil layers lies the parent rock, which is the main source of soil.

← **Earthworms are among the most visible forms of animal life** in the soil. Their feeding breaks down organic matter, and their movement through the soil creates small tunnels and tracks which help oxygen to be transmitted.

→ **Fungi of all shapes and sizes** live in soils and feed on plant and animal debris. Some species are associated with particular trees and plants, while others feed on a range of decaying matter.

↙ **In cool and temperate forests** dominated by deciduous trees, vast amounts of organic matter from leaves falls to the forest floor during fall and winter. In the deciduous temperate forests, this is gradually reduced to humus by the action of soil organisms.

↓ **Every 17 years cicadas emerge from the soil** under trees to swarm and mate. Once they die, the nutrients from their decaying bodies return to the soil.

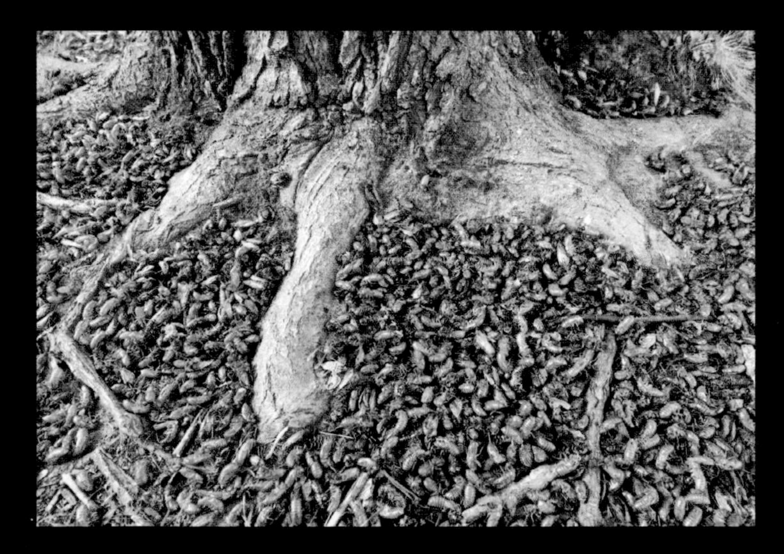

LIFE IN THE SOIL

The makeup of soil is constantly changing as it receives organic matter from trees and other plants and is eroded by wind and water. A large oak tree can deliver as many as 250,000 leaves to the soil when it sheds its foliage in fall. Many soils are teeming with animal and insect life, along with bacteria and fungi. Some soil-dwelling creatures, like earthworms, ants and beetles, are visible, while others, including bacteria and nematodes (roundworms), can only be seen through a microscope. In a constant cycle of renewal, the nutrients from trees and animals that live in a particular region is broken down in the soil and then taken up by the trees and other plants through the roots, to fuel their growth, and are then returned to the soil as debris to be broken down again by the soil organisms. The soils under deciduous trees that drop leaf litter, may be richer in nutrients than the soil under evergreen needle-leaf trees.

Leaves

Although leaves form different shapes and sizes, true leaves share a basic structure. Leaves arise from growth points on the stem called nodes. Most leaves are attached to a stem by a thin stalk called a petiole. The flat part of the leaf is known as the lamina. The shape and arrangement of leaves on the branch of a tree allow them to absorb sunlight, which is vital to growth. Leaves in all their myriad of shapes, sizes and colors have one main function that's unique to plant life. They manufacture food through a process called photosynthesis. Chlorophyll, the substance that makes leaves appear green, can turn the energy of the sun into sugars. These sugars are carried from the leaves to all sections of the tree, via vascular tissue known as the phloem. When trees are lacking moisture, their leaves respond by wilting. Veins in the leaves conduct the sugars produced by photosynthesis as well as transporting moisture. Despite their common function and structure, leaves vary enormously in appearance and arrangement. Some are single units known as simple leaves, while others, called compound leaves, are made up of many small parts called leaflets. The variety in leaves reflects the climate and conditions to which the tree is naturally adapted, and are used in identification and classification by botanists.

→ **The interconnected veins** seen within this leaf transport moisture and nutrients to other parts of the plant. The main vein, usually seen as an axis through the center of the leaf and traveling its length, is called the midrib. The system of veins is particularly visible in many leaves when they are held up to the light.

↑ **Oak leaves are discarded on the forest floor** during fall. The edge of the leaf, or the leaf margin, takes on many guises. It can be smooth, as it is here, edged with small hairs (known as ciliate), or points which range from small and toothlike (denticulate) to large and jagged (serrate).

↑ **The diversity of tree leaves** is evident in this comparison of the leaves of a conifer and a deciduous tree. The narrow, evergreen leaves of the Douglas Fir (left) survive winter protected under a waxy cuticle. The broad leaf of the deciduous maple is discarded in fall.

LEAF ADAPTATIONS

Leaves release water and gases, mainly oxygen and carbon dioxide, into the atmosphere through a process called transpiration. In cool or temperate climates, when the tree's roots receive adequate moisture, transpiration occurs without restriction. In hot or dry climates, systems within the tree reduce water loss to conserve moisture. One such method of moisture reduction is when the tree dispenses with true leaves and instead forms other leaflike structures. In some trees the petiole (the leaf stalk) thickens to become a leaflike phyllode. Where the stem has flattened to form a leafy structure it is referred to as a cladode. In dry areas, leaves may be thick and tough in an effort to conserve water. Referred to as sclerophylly, this occurs when sclerenchyma, the leaf's supporting tissue, is produced in large amounts. Other water-conserving adaptations in leaves include a waxy or hairy surface or an ability to droop to reduce the leaf's exposure to sunlight.

← **The sharp, scalelike leaves** of the Monkey Puzzle are arranged tightly around the stem and have a sharp, protective tip. The leaves can withstand frost and snow.

Leaves continued

The behavior of leaves divides trees into two distinct groups. Trees that have green, functional leaves through more than one season are known as evergreens, while those that lose their leaves due to seasonal changes are known as deciduous. In cool and temperate climates, deciduous trees begin their annual leaf loss in fall. In these climates, where winter is cold, often with many months of freezing temperatures, leaves are lost as the trees become dormant. In tropical and subtropical climates which experience distinct wet and dry seasons, trees may lose their leaves at the end of the wet season and remain bare and dormant through the dry. As trees prepare to undergo their annual leaf drop, they withdraw chlorophyll and nutrients from their leaves. Without the chlorophyll—which makes leaves appear green—other pigments within the leaf become apparent and color the leaves in tones of yellow, red and orange. Once all the nourishment has been withdrawn from the leaf, the stem is abscised and the leaf falls to the ground. When temperatures begin to rise, dormant buds begin to swell, making way for a new period of leaf growth in spring.

LEAVES AND CLIMATE

In broad terms, the size and shape of individual leaves is determined by the climate in which the tree grows. In moist, tropical climates, tree leaves tend to be large and robust in order to withstand hot sun and prolonged rain. Some leaves, such as those of the Sacred Fig (*Ficus religiosa*), have distinctive long tips, known as drip tips, which allow rain water to drain quickly from the leaf surface during storms. In hot, arid climates, leaves may be small and leathery, like the leaves of the Olive (*Olea europaea*), or sharp and spiky, such as the leaves of the Dragon Tree (*Dracaena draco*). In very cold climates with long, freezing winters, trees that remain evergreen generally survive by having small, tough needlelike leaves that are protected by a waxy cuticle. These leaves are arranged to closely hug each twig.

Spring For deciduous trees spring is a time of awakening. Many trees burst into flower, often blooming on bare wood. The new leaf growth is usually bright green in color but in some species, such as maple and ash, new growth may mimic the hues of fall, with gold or burgundy growth.

Summer During summer, as the deciduous tree's leaves reach full size, they work hard to manufacture food to fuel tree growth. Photosynthesis is most rapid on warm, sunny days when there is plenty of light, as well as moisture both in the soil and in the atmosphere.

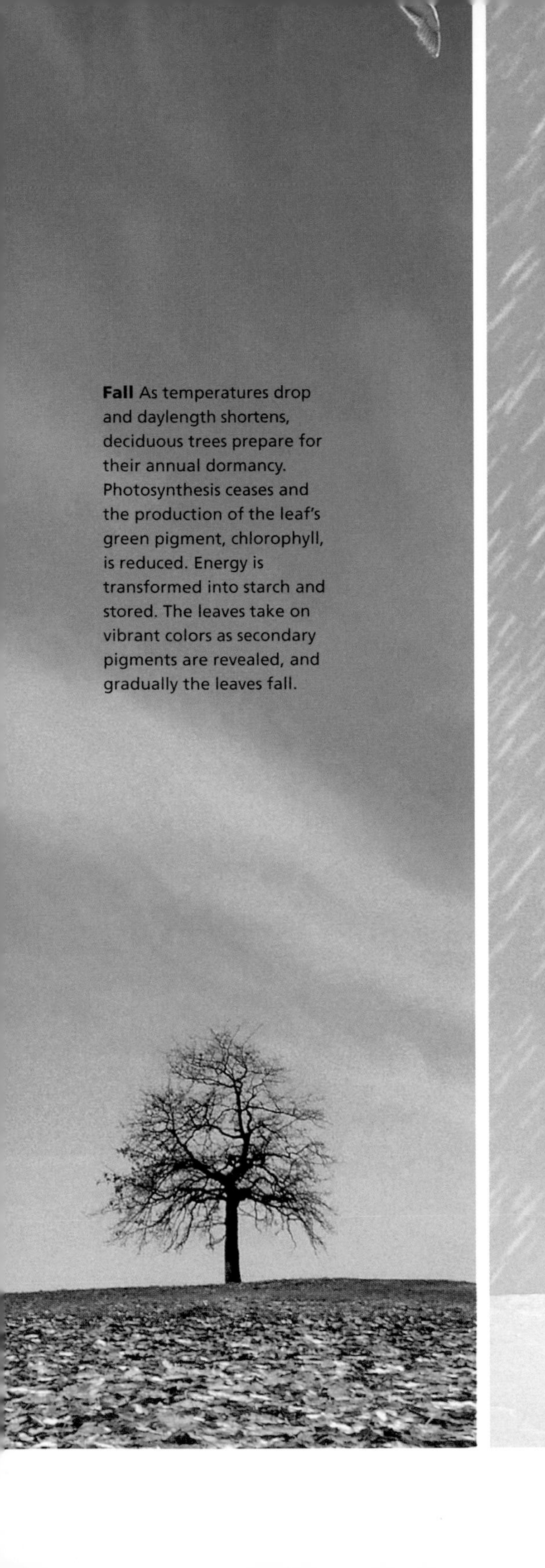

Fall As temperatures drop and daylength shortens, deciduous trees prepare for their annual dormancy. Photosynthesis ceases and the production of the leaf's green pigment, chlorophyll, is reduced. Energy is transformed into starch and stored. The leaves take on vibrant colors as secondary pigments are revealed, and gradually the leaves fall.

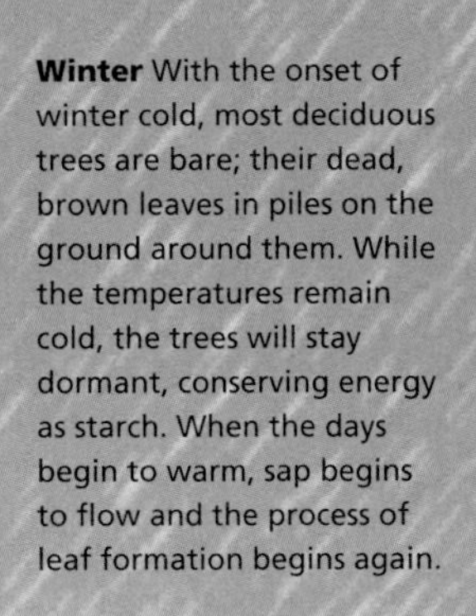

Winter With the onset of winter cold, most deciduous trees are bare; their dead, brown leaves in piles on the ground around them. While the temperatures remain cold, the trees will stay dormant, conserving energy as starch. When the days begin to warm, sap begins to flow and the process of leaf formation begins again.

↑ **The leaves of trees in wet climates** can have pendulous tips at the ends of leaves, known as drip tips, that allow water to run off.

↓ **Although this structure functions as a leaf,** the presence of a flower in its center indicates it is actually a flattened stem, known as a cladode.

↓↓ **The stiff, gray green "leaves"** of the wattle are leaflike in appearance, but are modified stalks, called phyllodes.

Leaves continued

Photosynthesis literally means light synthesis and refers to a plant's ability to produce food in the form of glucose, under the influence of sunlight, using carbon dioxide and water. During photosynthesis, oxygen is emitted as a waste product while carbon dioxide is absorbed and broken down. As well as being necessary to create food for plants, the process of photosynthesis is vital to all life-forms. Animals consume plants or plant-eating animals for energy and also breathe the oxygen plants emit. Animals in turn give off carbon dioxide as a waste product, which plants require in order to form glucose. Photosynthesis occurs in two stages. In the first stage, energy from sunlight is captured to form high-energy molecules. In the second stage these molecules transform carbon dioxide into glucose. Chlorophyll, a pigment that causes leaves to appear green, makes it possible for the conversion of sunlight into sugar. The chlorophyll can be found in the thykaloids inside the chloroplasts.

↑ **Photosynthesis** is a biological process that lies at the base of all life on Earth. Without plants' ability to turn sunlight into energy neither plants nor animals would survive.

↓ **The stomata,** through which carbon dioxide is absorbed and oxygen and water emitted from a leaf, are visible as small holes. Here the stomata of a eucalyptus leaf are shown magnifed by a colored scanning electron microscope.

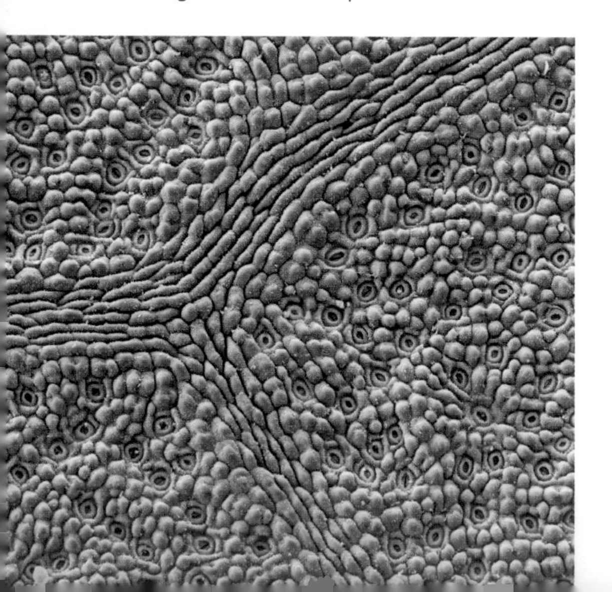

CHLOROPLASTS

Photosynthesis takes place in specialized structures within leaf cells, called chloroplasts. Chloroplasts are present in all the green parts of a plant, including the stem and buds, but it is in the leaves that the work occurs to capture the sun's energy and create sugars. Chloroplasts are shaped like small, flat disks, and are around 2–10 micrometers in diameter and 1 micrometer thick. Inside a chloroplast is liquid material known as the stroma and within this are the thykaloids (containing green chlorophyll), arranged in stacks, in which photosynthesis takes place.

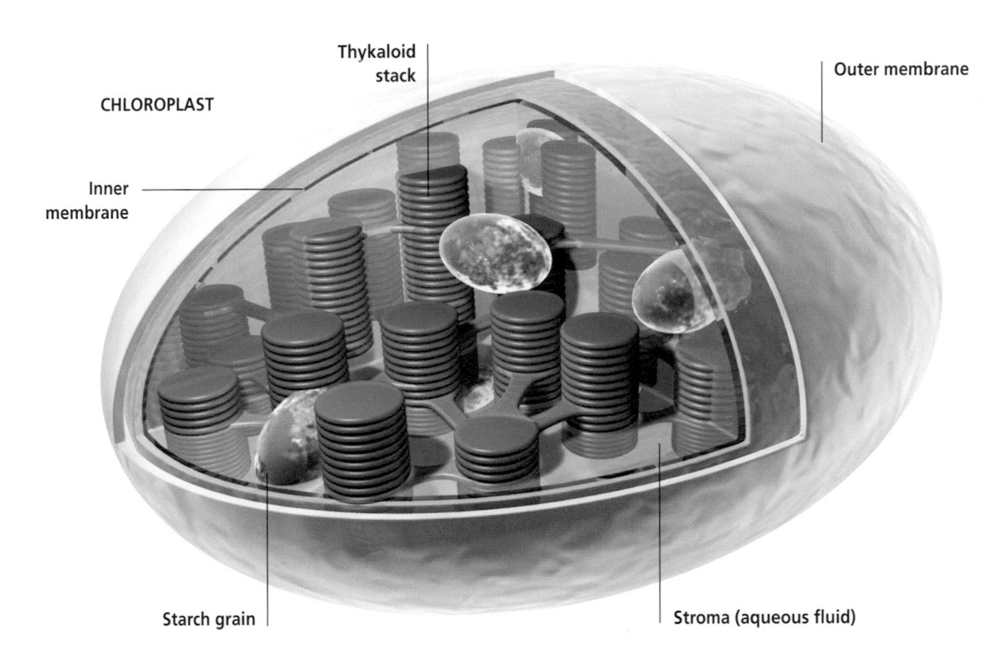

↓ **Leaf arrangements maximize** exposure to sun and minimize self-shading, as in this elm tree. Chlorophyll makes leaves appear green as it absorbs light strongly in the red and blue sections of the spectrum, but poorly in the green section.

Fertilization and reproduction

The purpose of a flower is not to appeal to our human senses but to attract insects and animals in an effort to spread its pollen to other, receptive flowers. Pollen is the male element of the floral reproductive system, akin to sperm in animals. Each grain of pollen contains a gamete. In order to fertilize the female floral parts to form seeds and eventually a new plant, pollen must move from the anther where it is produced. Since a plant is rooted in the ground, and its flowers firmly attached to branches, it can't move to find a mate. So the plant uses other vectors to allow its male and female parts to unite. The shape, scent, coloring and even position of the flower on the stem have evolved to increase each plant's chances of spreading its own pollen or attracting that of a mate. In the simplest flower structure, pollen from the anthers drops onto a receptive female stigma in a process called self-fertilization. However most plants, particularly trees, are designed so that self-fertilization does not occur. For seeds and fruit to form, most trees need cross-fertilization (pollen from another flower) and the resulting injection of new genetic material leads to seed formation.

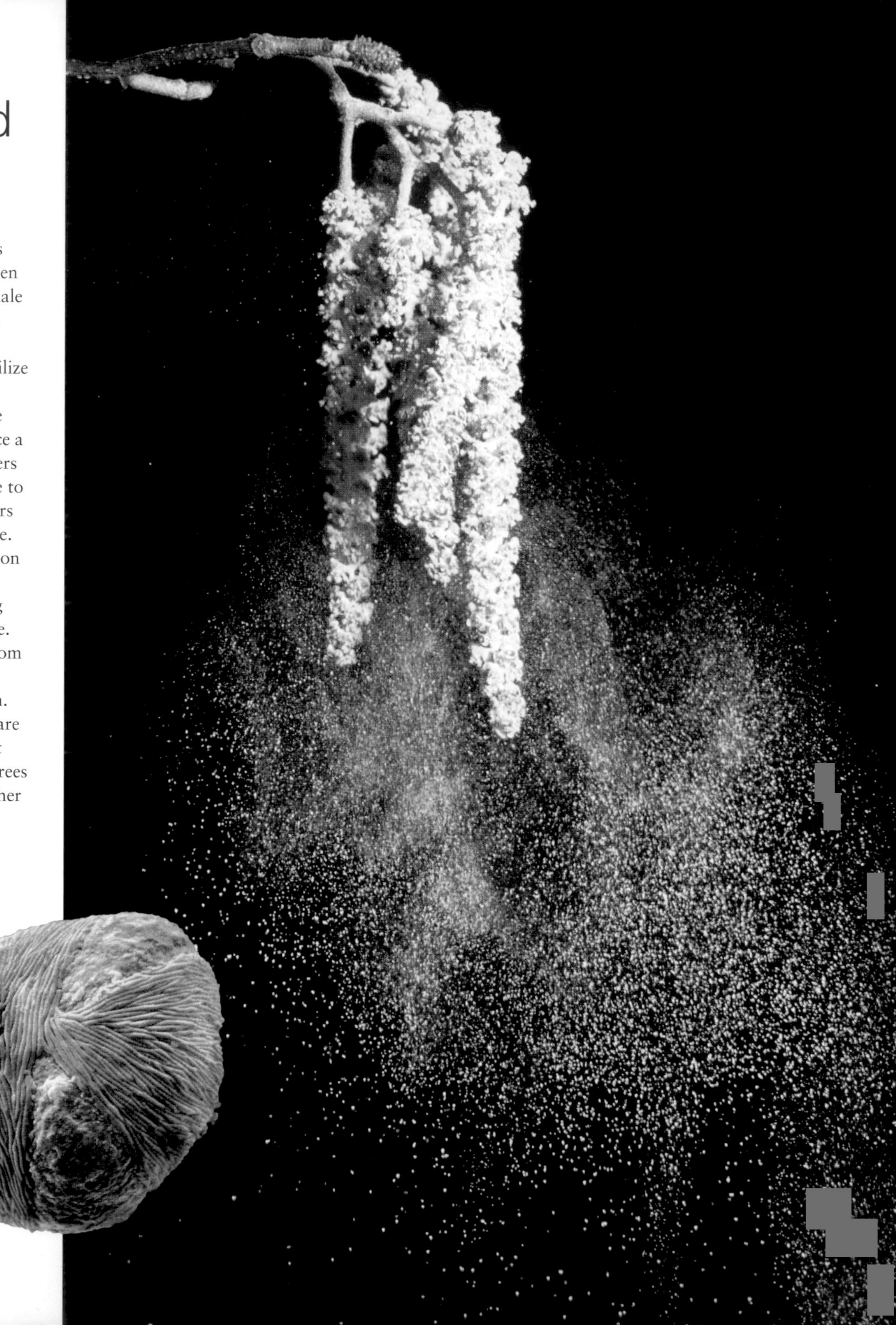

→ **A shower of pollen** becomes airborne as a moderate wind buffets the male catkin blooms of an Italian Alder (*Alnus cordata*).

→→ **The size and shape of pollen** varies from species to species. This is a pollen grain from a Sycamore tree (*Acer pseudoplatanus*), magnified 2000 times.

VECTORS OF POLLINATION

Trees use an array of methods to spread their pollen. Many rely on the wind to rustle through their flowers to release and spread pollen. Pollen enters the air and is carried far and wide. This is an inefficient method of spreading pollen since much of it fails to reach a receptive female, but instead falls on leaves, the ground or into water. Trees that rely on wind pollination, such as many conifers and broadleaf deciduous trees like elms and alders, produce vast quantities of pollen. Their flowers are usually small and insignificant. Often pollen is released in such large amounts by trees that it forms clouds that can cause allergies such as hay fever. Other flowering trees may rely on insects, birds or small mammals, such as bats or possums, to carry pollen from flower to flower and tree to tree, and also to spread their seeds in their droppings.

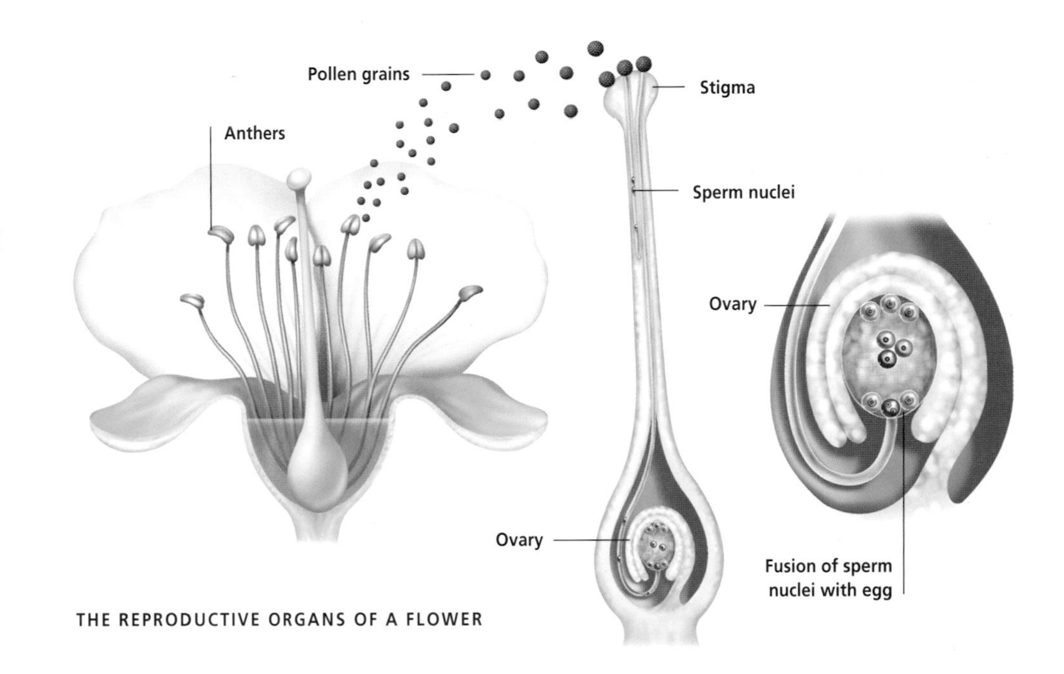

THE REPRODUCTIVE ORGANS OF A FLOWER

↓ **Birds spread tree seeds far and wide.** They also pollinate flowers by triggering a pollen release, which is then transferred to another flower.

↓ **Bats and other small mammals** may visit fruit trees, such as this mango, drawn to the strong scent, and then spread the seeds via droppings.

↓ **Many types of insects pollinate flowers.** Bees are active in daylight but some plants are pollinated at night by moths or beetles.

Flowers

Flowers carry the organs of plant reproduction. The male pollen is held on structures called anthers, and is received onto a receptive stigma, the female part of the flower. The stigma may be sticky or feathery in texture. Some trees have flowers that contain both male and female parts (hermaphroditic); others have exclusively male or female parts (dioecious); while others have separate male and female flowers on one plant (monoecious). These male flowers are called staminate while female flowers are termed pistillate. Many trees opt for simple, wind-pollinated flowers but others have blooms that attract visiting pollinators such as insects, birds or small animals. Flowers can become highly specialized in their shape or coloring so that just one species is lured to visit and pollinate. Other plants attract a wide variety of pollinators. Some flowers have developed a specific shape or structure that makes their nectar only accessible to insects or birds whose particular shape, weight or method of feeding suits the flower. Flowers with gaudy red or orange petals are often bird pollinated, since birds are attracted to these bold colors.

→ **The blossoms on eucalyptus trees** have dispensed with petals in favor of masses of pollen-coated anthers around the female stigma. In the flower's center a well of nectar attracts insects, animals and birds.

↘ **In order to reach the nectar** of Jacaranda (*Jacaranda mimosifolia*) flowers, bees must crawl into the blooms. At the flower entrance the bees brush against the anthers picking up a dusting of pollen.

↓ **The large, orange-red flowers of the Tulip Tree** (*Spathodea campanulata*) welcome visiting birds. The flowers are robust enough to withstand the birds' weight as they search for the nectar in the base of the flower.

↑ **The blooms of the Flowering Dogwood** (*Cornus florida*) are small and greenish in color, surrounded by large, showy pink or white bracts, around 1½ inches (3 cm) long.

INSECT POLLINATION

Insect-pollinated flowers share characteristics such as flower color (insects are attracted to flowers in blue and purple shades) and specific patterning or markings on their petals that are often visible only under ultraviolet light. These patterns may direct the insect to the flower's nectar. Insect-pollinated flowers may be fragrant and provide copious amounts of food in the form of nectar or pollen. Insects visiting tubular flowers must have a long proboscis (an elongated feeding organ) in order to reach the well of nectar within the flower or be small enough to crawl inside the flower. As they probe the flower searching for nectar or pollen, the insects come in contact with the anthers. Grains of pollen are released that adhere to the insect's body. The pollen is then carried to the next flower where it may rub off on a sticky stigma, the female receptor within a flower.

Flowers continued

Over millions of years of evolution, flowers have developed different shapes and petal arrangements. Flowers can be flat and open with many petals in a daisy-like shape known as rotate, or their petals may be fused together to form a pea shape. Other familiar shapes include tubular and trumpet. Flowers may be held on a single stem (a solitary inflorescence) or grouped together as a head of flowers, such as an umbel or cyme. Although flowers vary in shape and size, most have a common structure. A flower is a modified stem—a shoot whose leaves have become specialized as petals. Flowers open from a bud that is protected by a covering, the calyx, and made up of several sepals. The calyx is usually green. As the bud opens the petals—which, as a group, are referred to as the corolla—unfold to reveal to pollinators the sexual parts of the flower. The male filaments and anthers (the stamens) produce and hold pollen, while the female stigma and style receive the pollen so the ovaries can form seeds. Once the flower is fertilized, its life is over. The petals wither or the bloom may be discarded, leaving behind the carpel to swell as the seeds develop. This process is seen clearly in fruit trees, such as cherries.

FLOWERING AND NON-FLOWERING PLANTS
The variety of flower shapes, types of inflorescences and the arrangement of the floral parts within each bloom form the basis of plant identification and naming. Indeed, one of the basic groupings for plants is that of flowering plants (angiosperms), which includes flowering trees, shrubs, perennials and grasses; and non-flowering plants (gymnosperms), which includes conifers and cycads. Botanists date the appearance of the first flowering plants to the Cretaceous period, about 135 million years ago. One of the simplest of all floral structures is that of the magnolia, which was one of the earliest flowering plants to appear on Earth. Fossil records point to the existence of magnolia-like flowers, called *Archaeanthus*, more than 100 million years ago.

← **The Laburnum** (*Laburnum anagyroides*) is an ornamental but its seeds are poisonous. Its hanging racemes, or flower clusters, are made up of many yellow, pea-shaped flowers.

→ **The magnolia** has large, stiff, cuplike petals arranged around a center that is made up of fused anthers. It is thought that the magnolia bloom evolved in a shape that allowed beetles to crawl over the petals, thereby pollinating the flower.

← **Frangipani (*Plumeria rubra*)** has simple, whorled flowers described as funnel-form, each with five separate petals. The flowers are held in clusters called cymes. These subtropical flowers are usually white, often with a yellow center, but range from red to pale pink. They are perfumed and attractive to moths.

→ **The Judas Tree (*Cercis siliquastrum*)** has short racemes of pea-shaped flowers. The flower clusters are held on short stems arising directly from the axils of older branches and appear before the leaves.

Flowers continued

The extreme variety of flower types that has evolved over the past 135 million years has arisen in response to the many different methods of pollination. For instance, the position of flowers on the tree can determine the kind of pollinator. Blooms suspended on long stems, such as those of the Sausage Tree (*Kigelia africana*), are accessible to flying creatures like birds or bats, while those that cling to the trunk, such as the flowers of rain forest trees like the Cannonball Tree (*Couroupita guianensis*), may attract insects or small animals. The need to attract external pollinators is the flower's way of ensuring that it receives genetic information from other members of its species in order to increase the genetic variability of its progeny. As a survival strategy it can be a gamble, since without a pollinator the tree fails to set seed. The flowers of plants that can self-pollinate are guaranteed to produce seed with each flowering, but they limit their chances of genetic variability. Greater genetic variability means greater variety in each generation and therefore better chance of species survival.

FLOWER FRAGRANCE

As well as through the development of colorful petals, flowering trees often attract pollinators to their blooms with an enticing scent. Floral scents can range from the appealing (to the human sense of smell, that is) to the downright repugnant. Bees, butterflies and moths appear to visit flowers with "pleasant" scents. The fragrance of the Ylang Ylang (*Cananga odorata*) is incredibly sweet and is used in some of the world's most famous perfumes, including Chanel No. 5. Some flower fragrances are stronger at night, in order to attract nocturnal creatures, such as moths, and in many cases these flowers will be light-colored so they can be seen in the dark. Floral scent is a complex package of volatile organic substances, with up to 100 different chemicals involved in just one plant's perfume. Scientists have identified more than 1000 compounds given off by petals and other plant parts to form scent. As well as attracting pollinators, flower scents may also protect trees from predators and pathogens in the environment.

↑ **The weird flowers and fruits of the Baobab** (*Adansonia digitata*) are suspended on long stems and are pollinated by nocturnal animals, such as bats. One Madagascan species, *A. grandidieri*, is pollinated by lemurs.

→ **Witch Hazel (*Hamamelis* spp.)** has spidery flowers which are formed by long petals and contrasting sepals. Blooming in the cold of late winter, through the use of intoxicating scent, these flowers attract the attention of pollinating insects.

← **The catkin flowers of the Kilmarnock Willow** (*Salix caprea* 'Pendula'), lack petals but have many anthers, each with a dusting of yellow pollen. Willows flower in early spring before their leaves shoot. The pollen is spread by wind.

↞ **This rain forest tree from Ecuador,** *Herrania balaensis*, sprouts its red flowers directly from the trunk rather than from its extremities such as twigs or branches. This type of blooming is known as cauliflorous.

Cones

The gymnosperms are seed-bearing plants that include conifers and the palmlike cycads. Conifers, such as spruce, larch, pine and juniper, produce naked seeds that are held in cones, rather than the protected seeds associated with the other group of flowering plants, the angiosperms. Instead of producing floral structures, conifers produce male and female cones on the same tree. Male cones, made up of a stem and scales, bear pollen sacs. Pollen grains are dispersed from the sacs by the wind. The female cones carry ovules, which produce an embryo sac. Fertilization occurs once pollen from the male cone, carried by the wind, enters the female ovule, in what can be a long, slow process. In pines, for example, it may take more than 12 months from when the pollen reaches the female cone for fertilization to occur. Once fertilized seed forms deep within the cone, it then begins to mature. A ripe female cone is usually brown and woody however the size, shape and appearance of cones vary from one genus to another.

TYPES OF CONES

Cone scales are actually tightly packed leaves known as sporophylls. Cones can vary in size from the small button-size cones of junipers to large cones the size of footballs, such as those of the Bunya Bunya Pine (*Araucaria bidwillii*), which can be 1 foot (30 cm) across and may weigh over 2 pounds (1 kg). The seeds inside the Bunya Bunya cones are starchy and highly nutritious and were prized by Australian indigenous people as a protein-rich food source. When these large cones fall, they usually shatter, spilling the seeds. Other conifer cones, particularly those of pines, are also harvested for the nutritional value of their seeds.

← **The Sandplain Cypress** (*Actinostrobus arenarius*) belongs to a genus that is endemic to Western Australia. The small male cones are around ¼ inch (5 mm) long and are located on the tips of the twigs. The female cones (left) are slightly larger at ½–1 inch (10–20 mm) long and are globular to ovoid in shape, with six thicky woody scales which are closed when immature, but open out slightly when mature.

↑ **This immature larch cone** has tightly overlapping scales. When young, the cones are green or purplish and take approximately 5–8 months to reach maturity after fertilization has taken place.

→ **Mature larch cones are upright** and ripen in summer but can be held on the tree for many months. These cones are persisting on the branch even as the leaves begin to color with the approach of fall.

← **The process of fertilization** is slow in all conifers but particularly in pines. The female pine cone may take 12 months or more to change from the immature state (far left) to the mature pine cone (left). Once the pollen reaches the female cone it enters the ovule where fertilization takes place.

Fruits and nuts

Fruit is the ripened ovary, complete with seeds, of a flowering plant. Fruits may also contain both the ovary and some surrounding tissue. Seed is formed in the female section of the flower, known as the carpel. Where a plant produces separate male and female flowers, only the female flower produces seeds and fruits. In most plants, the floral parts—petals, stamens and anthers—are discarded once the seeds form. Sometimes the petals or sepals may remain attached to the developing fruits. Their remnants may be seen, for example, at the base of apples. In false fruits, other parts of the flower are contained inside the fruit. For instance, in an apple, the receptacle swells to form the fleshy part of the fruit and within this the carpel and seeds are found and form the core. In a true fruit, such as a plum or cherry, the carpel wall swells to form the fruit, or pericarp. True fruits include drupes (a fruit that contains a seed encased within a hard stone) and berries, which contain many seeds and result from several fused carpels. Interestingly, the orange, with its multiple seeds, is correctly classified as a berry. A fruit such as the pomegranate, which contains several seeds, each enclosed within a stone, is known as a compound fruit. In some trees the entire inflorescence, or head of flowers, becomes a single fruit. These are known as composite fruits and include figs and mulberries.

LONGITUDINAL SECTION OF A FIG

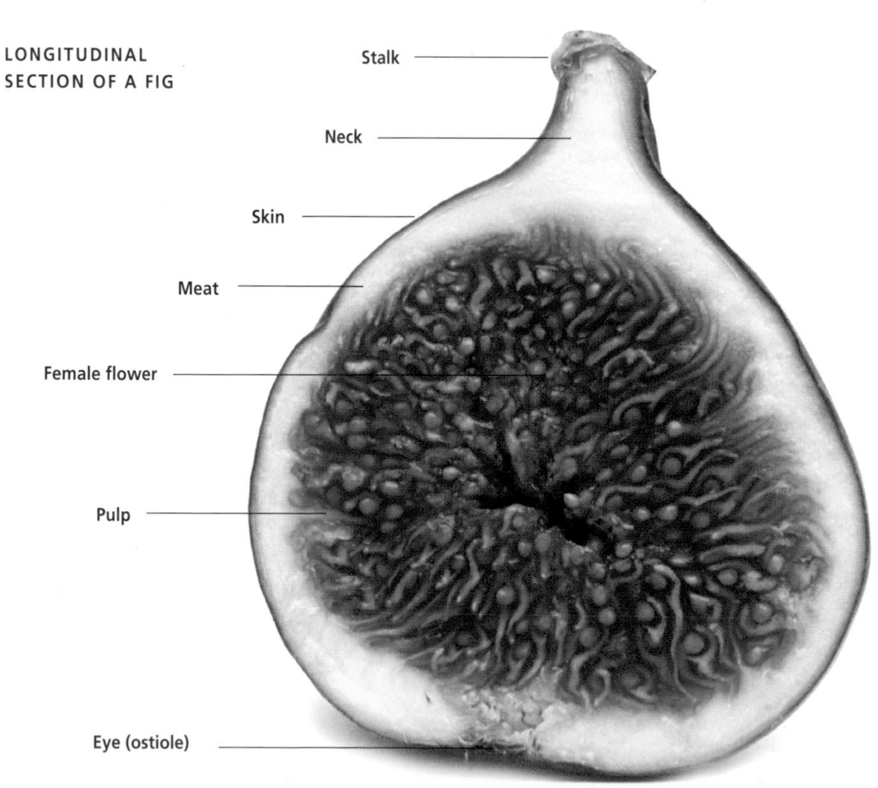

↑ **Spiky durians** are large, strongly scented tropical fruits. In Singapore they are banned from being carried on the public transportation system due to their offensive odor. The rough surface is covered with sharp thorns and the soft, five-segmented, pulpy fruit surrounds the seeds.

↗ **The huge 12-inch (30-cm) long pods of the Cacao** (*Theobroma cacao*) tree cling to the trunk and branches. The pods are harvested and then roasted to produce the basic ingredients of cocoa and chocolate.

→ **The fruit of the orange tree** (*Citrus sinensis*) is a hesperidium, which is actually a kind of modified berry (an edible fruit with multiple seeds). The flavor of oranges can vary from sweet to sour.

→→ **Plums provide a rich, sugary mass** of edible fruit around a single stone, or hard endocarp, containing the seed. A plum is a fruit type known as a drupe.

← **Figs are unusual fruits.** In most fruits the flowers, with their stamens and styles, are open to the air and pollinators. Once pollination occurs, the fruit forms and the floral parts are discarded. However, in figs the flowers are inside the fruit. Fertilization can only occur if a single wasp species, known as a fig wasp, can access the fruit.

NUTS

Most of the popular edible nuts are simple dry fruits where the ovary wall or pericarp of the flower forms a hard outer covering around the fertilized ovule. Unlike fruits, nuts do not split open at maturity to release their seed. Some commercially important edible nuts produced by trees include almond (*Prunus amygadalis*), walnut (*Juglans regia*), macadamia (*Macadamia integrifolia*), hazelnut (*Corylis avellana*) and chestnut (*Castanea sativa*). The edible, naked seeds of gymnosperms, which include conifers, are also sometimes referred to as nuts. These include pine nuts (from *Pinus* spp.) and ginkgo nuts (from *Ginkgo biloba*). Not all nutlike structures are nuts. For instance, the peanut is a legume, the coconut is actually a drupe, and Brazil nuts and cocoa pods are actually seeds.

↓ **Almonds** (*Prunus amygadalis* or *Prunus dulcis*) are related to the fruits of the *Prunus* genus, but the fleshy outer covering is replaced with a leathery coat.

Seeds

Not all seeds have the soft, edible surroundings provided by fruits such as mangoes, plums or apples. Some seeds are contained in a dry, woody and unpalatable case, also called a fruit. These receptacles vary in shape and size but all provide some type of protection for the seed they contain. This protection is vital for the survival of the tree as it germinates, grows and finally forms flowers and fruits of its own. Seeds are rich in starch and protein, both highly desirable sources of energy for animals. The woody or fibrous surroundings of a dry fruit have evolved to protect the seed from being eaten or damaged before it can germinate. Like soft fruits, dry fruits come in many different shapes and sizes. All can be divided into three main groups based on how the seed is released from within its woody case. A dry fruit that splits apart to release its seed is known as dehiscent. It may split explosively like a capsule or a pod on an acacia, or split down one side like the pod of a grevillea. Indehiscent seeds do not split open. Their seeds are not released until the container rots. In some plants the fruit splits but into sections that contain seeds. These are described as schizocarpic. Each section, known as a mericarp, must then rot down before releasing its seed. Dry fruit can vary in size from the small nuts of a hazelnut to the exceptionally large fruit of the coco de mer.

SEED FORMATION AND STRUCTURE

Seed, which carries the germplasm to grow a new plant, is formed when the ovule in the flower is fertilized. Within each single seed is the recognizable beginning of a plant, poised to grow and develop. A seed has a seed coat (called a testa), which surrounds the growing part or embryo and provides a food resource to help fuel growth. The embryo itself contains the root (known as the radicle), which grows downward, along with the stem (plumule) and seed leaves (cotyledons), which head upward to the light. The size of the fruit and seed have no relationship to the size of the tree that can develop. Tall eucalypts grow from the small fine seeds that spill from a gum nut while a small palm grows from the coco de mer seed, the largest seed produced by a living plant.

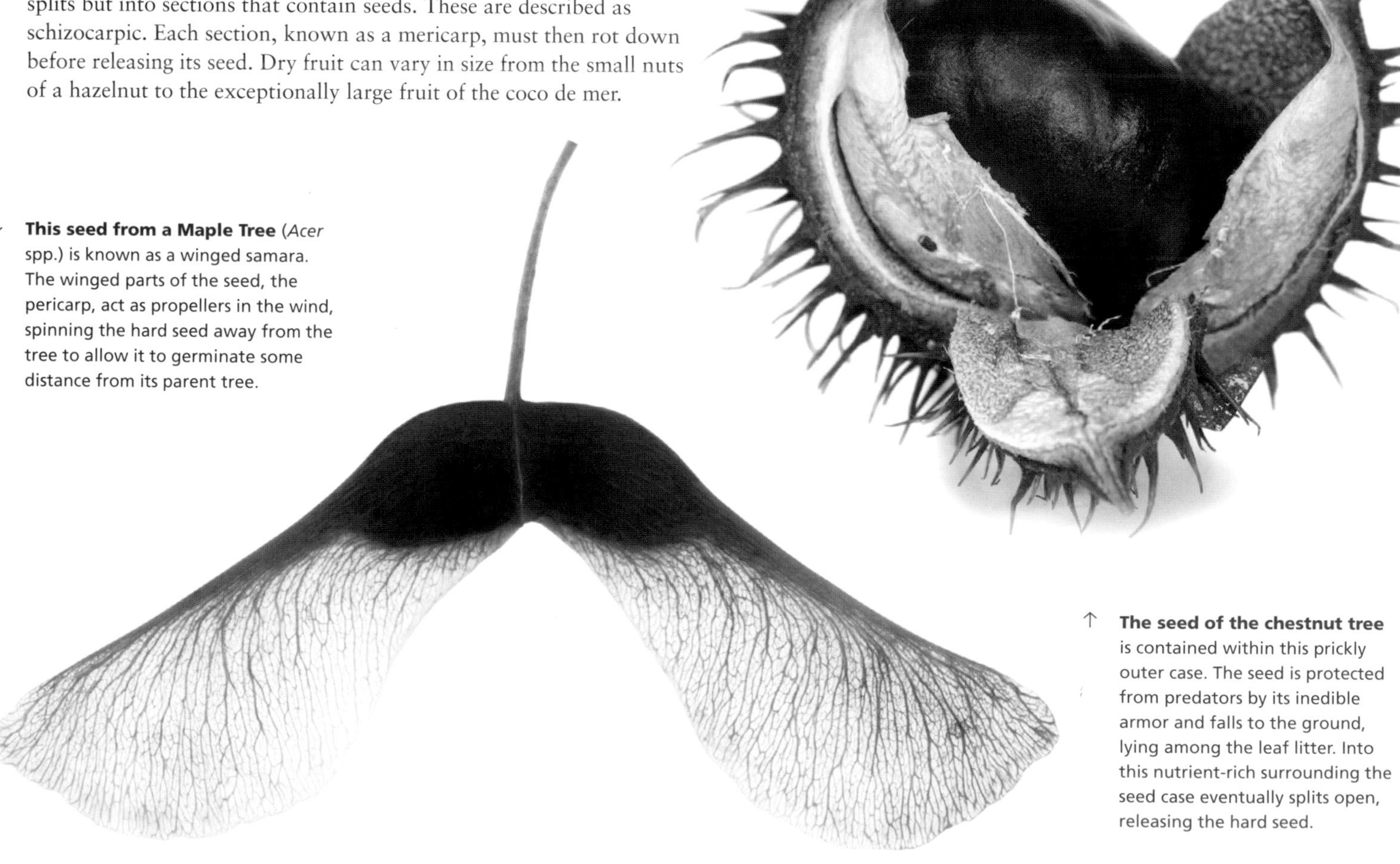

↓ **This seed from a Maple Tree** (*Acer* spp.) is known as a winged samara. The winged parts of the seed, the pericarp, act as propellers in the wind, spinning the hard seed away from the tree to allow it to germinate some distance from its parent tree.

↑ **The seed of the chestnut tree** is contained within this prickly outer case. The seed is protected from predators by its inedible armor and falls to the ground, lying among the leaf litter. Into this nutrient-rich surrounding the seed case eventually splits open, releasing the hard seed.

↑ **Once a cluster of eucalyptus flowers,** this is now a cluster of seed capsules. When mature it opens, allowing the seeds to spill out. In some environments, where fires are prevalent, the seed remains in the capsule until the heat or smoke from the fire triggers its release, ensuring germination.

← **Magnolias are among the oldest flowering trees.** Once the flower is fertilized, the seed forms into this unusual fruit that can remain on the plant for many months before it is discarded and its seeds released.

↓ **The seed of the coco de mer (*Lodoicea maldivica*)** palm is the largest seed in the plant kingdom. It is around 20 inches (50 cm) in diameter and can weigh up to 65 pounds (30 kg).

Tree shapes

A tree's shape is determined both by its genetic blueprint and by its environment. Trees come in a variety of shapes from the tall and narrow habit of many conifers to the broad and spreading form of an oak tree. For most trees, initial growth is upward in a pattern known as apical dominance. Growth hormones or regulators that are produced in the main growing tip of the tree, deter other growth from occurring elsewhere on the tree. This system enables trees to extend upward to seek light. Apical dominance is important to trees growing in a forest, where there is limited space and light. Trees with a naturally triangular shape, described as conical or pyramidal, continue to grow under the influence of the central growing point of the tree—the apex or main leader. Other trees soon lose the strong apical dominance that formed their early growth, and develop a strong, broad shape with horizontal branching. Interruption to the single upward growing point by physical damage, such as in a storm or by fire or insect attack, halts the flow of hormones from the apex, allowing other growing points on the tree to become more dominant, and may result in a tree forming multiple growing points, which over time leads to a new shape that has been molded by the environment.

DIVERSITY
A tree growing near the coast, that is constantly buffeted by salt-laden sea breezes and growing in sandy soil, may develop a low, spreading shape more like that of a shrub. This type of growth is often referred to as scrub or mallee growth. Further inland, with less wind exposure and access to deeper, more fertile soils, the same species may be a tall, stately tree with strong upward growth and outward-spreading branches. Because a single tree species may, in different growing situations, develop highly variable forms, tree shape is not always a reliable method of species identification.

← **Conifer forests** are characterized by trees with a regular, conical or spherical shape. They also exhibit strong apical dominance with a single, vertical growing point and horizontal branching patterns.

<< **Palms have a growth pattern** and shape that differs from other forest flora. Most palms lack side branches and grow upward with a crown of foliage. If the growing point is damaged, the tree dies. In some cases palms may form a cluster of trunks, each with its own crown.

← **The Umbrella Thorn Acacia** (*Acacia tortilis*) develops its characteristic, spreading crown only at maturity. Common in arid savanna areas of Africa, it is useful as a shade tree for both humans and animals.

<< **The weeping shape** of this willow is formed by a single trunk with many side branches which form weeping branchlets. Weeping trees are found in environments as diverse as moist riverside habitat and inland arid climates.

Growth stages

Every tree progresses through successive stages of growth. Beginning with the seed, the first stage is germination, which is the breaking of the embryo's dormancy and emergence of shoot and root, using the seed's stored food. Depending on how much food the seed contains, this energy source will last a shorter or longer time. The seedling stage comes next as the new roots begin getting nutrients from the soil; it may be very brief or may last years; height growth is slight but below ground a strong root system develops and the stem base and taproot thicken and strengthen. The seedling progresses, in some plants after several years but in others almost straight away, to the tree's fastest stage of growth, referred to in its earlier part as the sapling, but overall as the "grand phase of growth." This is where many trees gain nearly all of their height, and lasts from around 8–25 years. Some fast-growing trees can add 7–10 feet (2–3 m) annually in height. Growth tends to be upward, with a very strong apical shoot and lateral branches short. Some trees, mostly tropical and large-leafed, produce no lateral branches during this phase.

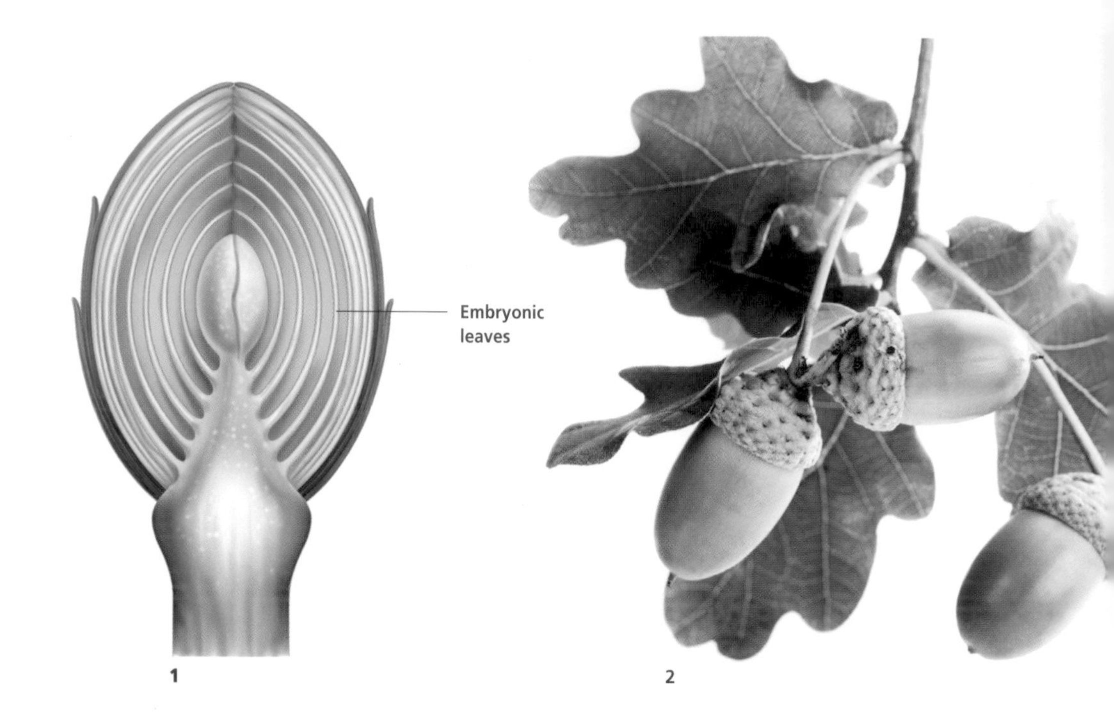

MATURITY

A tree entering the mature stage switches its energies to spreading its canopy wider, increasing its wood bulk, and producing large quantities of flowers and fruit (or cones in the case of conifers). Depending on tree type, it may last as little as 10 years (for example, some *Acacia* species), or up to hundreds of years (for example, the redwoods). Then follows a transition to the "overmature" stage, when the tree begins a long, slow decline. Height growth slows almost to a stop, though the canopy can still spread slowly, and annual growth rings in the wood become closer and closer. But growth of wood and shoots never actually stops, since that would be death in any plant. Finally there is senescence, with a decaying, hollow trunk and dead limbs, followed eventually by death. In the case of the Bristlecone Pine, senescence may last over 1000 years.

5

3

4

← **1. A tightly furled bud** containing embryonic leaves **2. Ripening acorns** of English Oak (*Quercus robur*). Each acorn is a leathery-skinned fruit enclosing a single large seed, which must overwinter on the ground among moist leaf litter to break dormancy. At the time of germination the stored nutrients give the seedling a strong start. **3. This rowan sapling** entering on its "grand phase of growth," predominantly upward. **4. A mature oak** with a broad, healthy canopy, but with annual height growth now small.

6

← **5. This oak tree is overmature,** with sparse canopy and trunk scarred by stubs of fallen branches. Growth still continues at ends of the long limbs, which droop lower every year. A tree like this may still produce heavy acorn crops. **6. A dying or dead tree** eventually topples to the forest floor. The process of decay begins, often with wood- and bark-eating insects, such as beetles, but also with bracket fungi and other "toadstools," whose wood-attacking mycelia riddle the whole trunk. The decay process ends with minerals and some carbon returned to the soil, and carbon dioxide returned to the atmosphere.

Tree growth

True wood is only produced by conifers and dicotyledons, and then only by those species whose aboveground growth is indefinite, that is, trees and shrubs. The features distinguishing true wood are that it is laid down steadily from the outside in continuous concentric layers by the dividing cambium cells; and that it consists of strong, lignified cells, most of which are able also to conduct water upward through the trunk while they are alive. In fact it is only the live outer zone of the wood, called the sapwood, that conducts; after a number of years its cells die and become blocked by tannin deposits and other materials, thus turning into heartwood. But for every year of heartwood that accumulates, a new layer is added to the sapwood, which thus remains a more-or-less constant thickness. Trees grow therefore in two ways: by elongation of their branch tips, with lateral branches added and also elongating; and by adding layers of wood, not only to the trunk but to all the limbs, steadily increasing the tree's bulk and mass. While wood cells are added by the inner face of the cambium, bark cells are being added by its outer face.

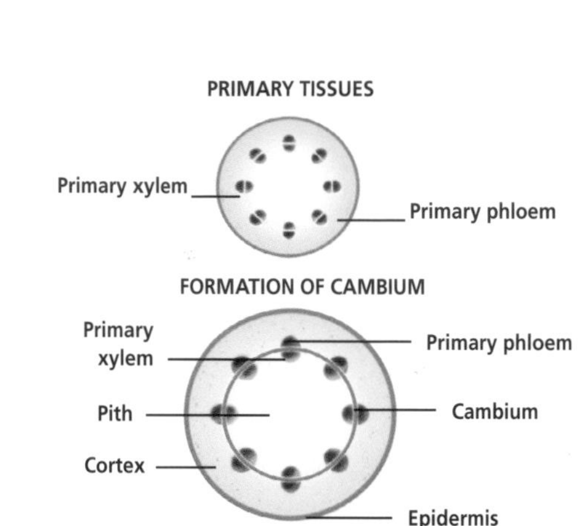

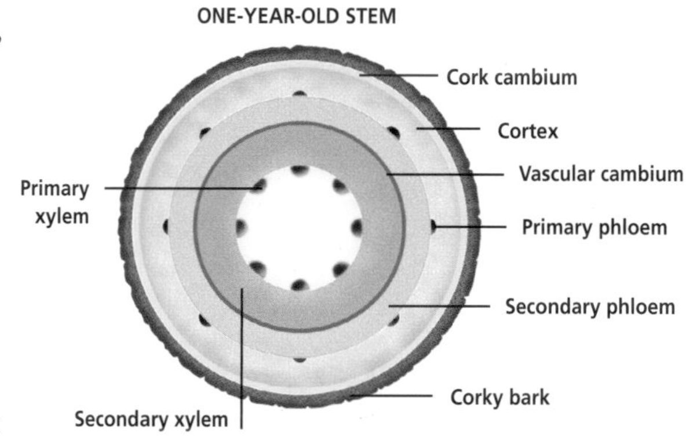

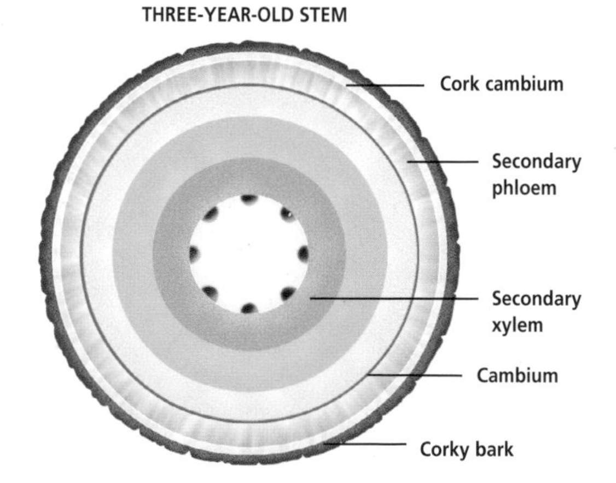

↑ **A diagram showing the development** of a tree stem and growth over a three-year period, showing the primary xylem and primary phloem and the gradual formation of the cambium, the secondary phloem and xylem and the bark.

ANNUAL RINGS

In many trees from tropical and subtropical climates, growth is almost continuous throughout the year, and so the wood layer added is of fairly even texture and cell size. But in trees from colder climes that enter into winter dormancy (whether deciduous or evergreen), wood growth likewise slows and almost stops by winter. At summer's end, when temperatures start dropping, the wood cells become smaller and their walls proportionately thicker. The large late-spring–early-summer cells make up what is called the early wood, and the diminishing late-summer and fall cells the late wood. Each early and late wood pair is called an annual ring. In some dicotyledons the early part of the season's growth is also marked by a zone of more crowded and larger vessels (pores); such woods are termed ring-porous. The wood of conifers, by contrast, lacks vessels altogether, while in other dicots they are scattered through the wood in many varying patterns.

↑ **In this historic photo,** tourists examine the growth rings of a felled 996-year-old *Sequoiadendron* tree.

← **Cross-section of a mature pine trunk,** showing about 50 annual rings. Each ring consists of a pale band of early wood and a thinner, darker band of late wood. Differing widths of rings reflect variations in rainfall or temperature in growing seasons.

↓ **Microscopic cross-section of one growth ring** of a conifer. Early wood, with large, thin-walled cells, occupies the bottom four-fifths; the top fifth is late wood, with smaller, thicker-walled cells. At the lower right a large resin canal can be seen.

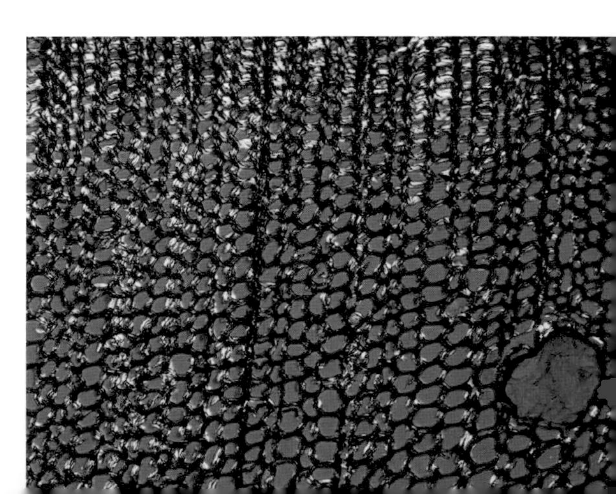

Attack and defense

Since plants are at the base of the food chain and immobile, they are at great risk of being eaten. Over millions of years of evolution, trees and other plants have evolved defense methods to protect them against all manner of predators, from pathogens to grazing animals. Many trees use physical barriers such as thorns, spines or prickles as their defense. Thorns on the trunk prevent larger animals from climbing the tree to eat the fruit or foliage. Thorns are sharp, woody structures formed from the stem. They are found on the trunks and branches of trees such as hawthorns, Bombax, citrus and Kapok trees. Other sharp deterrents are formed by other parts of the tree. Spines, found on some acacias, are modified stipules—a small, usually soft growth that occurs at the base of a leaf stalk. Prickles are outgrowths or emergences from the stem, rather than a modified plant part as with a true thorn. Because plants have evolved more sophisticated methods of protecting themselves from being eaten or damaged by pests, so too have insects and animals developed ways to avoid a plant's physical defenses while still getting access to the nutrients stored in a plant. They may be able reach the tree's food sources without climbing, thereby avoiding thorns, have small mouthparts that enable them to feed among spines, or tough mouthparts that are impervious to prickles.

→ **The spiny trunk of this tall rain forest tree,** the Floss Silk Tree (*Chorisia speciosa*), is far from welcoming. Although birds, moths and bats visit the crown of the tree and are its main pollinators, climbing animals that may eat its leaves find access impossible.

→ **Eucalyptus leaves** such as this colored and magnified section from *Eucalyptus citriodora*, seen through a scanning electron micrograph, contain lots of oil, which for most animals make the leaves inedible. The oil covers the leaf surface with layers of a white, crystalline wax.

← **These pairs of sharp, vicious spines** are modified leaf stipules protecting the native African tree, *Acacia karroo*, from being grazed. In many countries karroo thorn is an environmental weed.

→ **Resin is made in special channels** called resin ducts in a criss-cross pattern. The duct is lined with cells that form and release resin into the duct. If a tree is wounded or a branch snaps off, the resin oozes out. Once it is exposed to the air, it eventually solidifies to form a plug.

← **Holly is an evergreen tree** that not only retains its leaves though the coldest winter weather, but has also made itself unappealing as a winter food for predators. The sharp, spiny edges to these stiff, glossy leaves render them inedible.

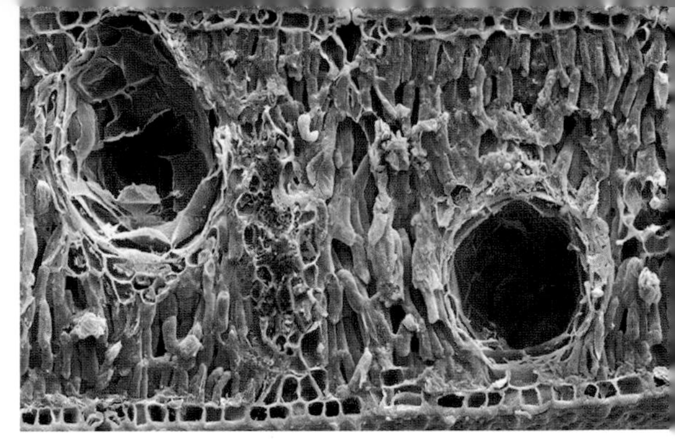

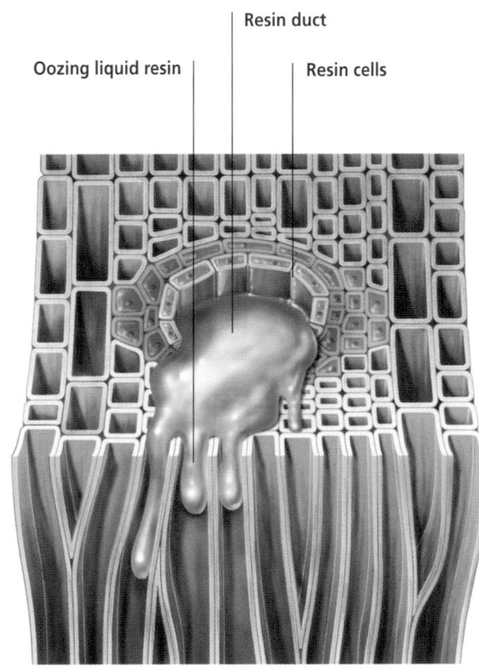

CHEMICAL DEFENSES

Some defenses are invisible. Trees may contain chemicals that cause skin irritation or even poisoning if parts of the plant are touched or consumed. Some leaves produce copious quantities of oils or tannins to make them unpalatable, such as the eucalyptus leaf, while others trigger irritating skin rashes if they are touched. Rhus Tree (*Toxicodendron succedaneum*) causes dermatitis in humans that come in contact with its leaves or stems. Healthy trees may repel insect attack by chemical means. Repeated feeding on some species appears to increase an individual's ability to protect itself. Studies have shown that new citrus foliage on plants that are constantly eaten by caterpillars becomes less palatable than foliage that has not been eaten.

Trees are preyed on by animals and insects of all shapes and sizes. Insects may be small, but they can completely defoliate a tree, and may also attack the buds, shoots and bark. Despite the tree's defenses, including thorns or spines, large, leaf-eating animals like elephants and giraffes may be undeterred. Disease organisms, including fungi, bacteria and viruses, may also attack any part of a tree. Somtimes, as a result of infection, small round galls are formed, which are swellings of plant tissue. Certain insects and fungi are necrophytes, which means they feed on dead or decaying wood or bark, while others feed on living tissue. The long-term effect of any pest or animal attack on living tissue varies with the type of problem and the age, health and vigor of the tree. In many instances, trees may be damaged but not killed. However, often the attack leads to the plant's death. Insects that enter the tree's sapwood may cause the tree to die by impeding the flow of water and nutrients through its vascular system. In many instances, insect attack alone is not enough to permanently harm or kill a tree. Secondary problems, such as disease, may enter wounds with drastic results. Storm, lightning or fire damage can also cause wounds which leave a tree at the mercy of infection from bacteria or fungus.

→ **Despite their enormous size,** elephants are herbivores and can eat up to 440 pounds (200 kg) of plant matter per day. This African elephant is feeding on an acacia, seemingly undaunted by the thorns.

↙ **These bright red lumps** on the leaves of a plane tree are galls, formed in response to a parasite. Galls can be caused by bacteria, fungi, nematodes, insects or mites, or by a combination of these.

↓ **The Dryad's Saddle Fungus** (*Polyporus squamosus*) is parasitic and eventually will kill the host tree.

↘ **This forest tent caterpillar** (*Malacosoma disstria*) is feeding on the leaves of a sweet gum. Healthy trees survive attack by growing new tissue or fighting back with chemical defenses.

PLANT ATTACK

As well as being attacked by animals, insects and birds, trees are often under siege from other plants, such as vines, parasitic growth like mistletoe, or simply by the encroaching plants surrounding them. Climbing vines may damage trees by smothering their leaves, thereby reducing or preventing photosynthesis from occurring. Vines are often weed plants that have been accidentally introduced into a forest ecosystem. If unchecked they may cause whole sections of the forest to decline and reduce the natural habitat. Some vines literally strangle a host tree by interfering with its flow of nutrients and water in its xylem and phloem. The Strangler Fig, for example, begins as a small plant living in a tree branch. Over time it grows larger, its aerial roots reaching the soil, until it eventually engulfs the host tree and itself becomes a massive forest tree. Mistletoe is another parasitic plant that grows on the branches of trees. A modified root penetrates the bark of the tree, allowing the mistletoe to take water and other nutrients from the host tree. If infestation becomes too heavy, the tree will die, but more commonly, tree growth is only stunted. Mistletoes are spread by birds, who eat the seeds and then spread them through their droppings.

Survival of the fittest

Trees face many obstacles as they develop and reproduce. Even at the beginning of a tree's existence, with germination of a seed, the odds may be stacked against it. The new seedling may find itself in soil too shallow or stony, or drying rapidly after brief rainfall. If its roots cannot quickly find moisture and nutrients, the seedling will die. At the other extreme, thousands of seeds may germinate in good, moist soil on one small patch under or near a parent tree. As they grow, they must soon compete with one another for nutrients, water and light. This is just the start of the long process of competition among trees in a forest, between different species or between trees of the same species. The winners are those that outpace others in height and then spread their canopies, robbing their slower companions of light. These are the trees that will in turn produce the largest quantities of seed, thus giving their genes a competitive edge. But competition is only one part of a tree's struggle. In most trees' native forests there have evolved a whole spectrum of organisms for whom the tree is a food source. Insects are perhaps the most numerous, chewing the leaves, mining through their tissues or sucking the sap, tunneling through the wood and inner bark, or beneath the soil eating the tree's roots. The tree must either make growth fast enough to outpace the insects' eating, or must have evolved defenses against them, for example flooding sites of attack with sticky resin, latex or tannin. But insects are far from being the only type of fauna that feed on trees.

NATURAL COMPETITION

At certain stages of growth trees may also face competition from other kinds of plants—dense weed growth can smother seedlings or divert water and nutrients from older trees; vines can smother trees' canopies, depriving them of light. Parasites, including other green plants such as mistletoes, can rob food from the trees' sap. Some tree species, such as walnuts, defend themselves against competing plants by poisoning the soil with their leaf litter. Most mature trees grow in precarious balance with destructive microorganisms, such as some fungi, which are continually killing or weakening their internal tissues, especially their roots. Then there is a whole range of physical challenges that trees must face—hurricanes, salt spray, soil salinity, landslips, fire, flood, drought, heavy snowfall and frozen soil. Trees have evolved adaptations to better cope with all of these, and the best-adapted species have a competitive advantage over other trees.

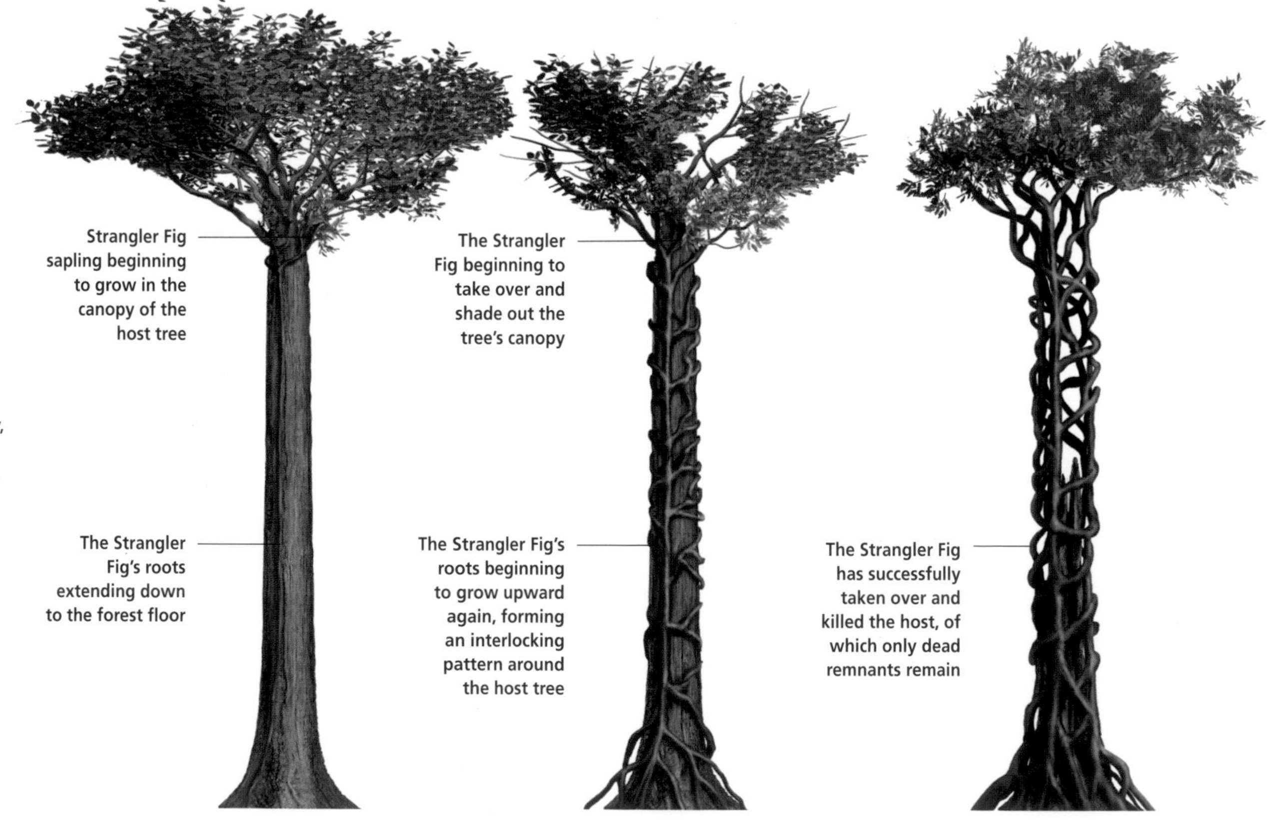

→ **The Strangler Fig** is a tree that begins life high in the branches of another, host tree, after one of its seeds is dropped by a visiting bird. Over time the fig sapling grows aggressively, extending its roots down to make contact with the ground. The roots then begin to form in an interlocking pattern around the trunk of the host tree, eventually strangling and killing the host by cutting off its flow of nutrients.

← **Woodland in southeastern** USA being overwhelmed by the Kudzu Vine (*Pueraria lobata*), a beanlike legume from East Asia, originally introduced for fodder and erosion control. As well as killing trees by robbing of light and soil moisture, the dense growth of the vine prevents regeneration of seedlings. The trees' only defense may be to wait for decades until the vine's vigor declines and they can regenerate from soil-stored seed.

← **In an overhead view** of a Central American rain forest canopy, one can see how a single specimen has vigorously grown above its surrounding competitor trees.

Wind and weather

Trees are all-weather organisms. Apart from their obvious need to adapt to summer heat and winter cold, they are nearly all, to a greater or lesser degree, adapted to withstand strong wind. If you think of how quickly an umbrella or a loose boat sail is tattered and destroyed in a gale, it seems quite miraculous the way tree foliage can survive under the same conditions. The anatomical structures of leaves, stalks, twigs and branches, as well as their external architecture, are highly adapted to allow flexing and streaming in the wind without suffering the sort of mechanical fatigue that ends in fracture. A prime example is a palm frond, such as that of a Coconut Palm, which can be seen bent backward and streaming in a hurricane that may last for days, afterward returning to its normal form virtually undamaged. And it is not only the foliage of these and other trees that are built to survive gales, but their trunks and roots as well, providing support and anchorage that combines stiffness with resilience. Naturally the highest resistance to strong winds is found in trees that are frequently exposed to them, for example many conifers from high mountains, which have flexible, almost rubbery, twigs and tough, needle-like leaves. These same features allow them to better withstand hail, which is prevalent in many mountain regions and inflicts greater damage on broadleaf tree foliage (though even that can be quite hail-tolerant). And then there is snow. It is well known among foresters, farmers and gardeners in regions of snowy winters that some tree species are badly damaged by a heavy snowfall, their branches and leading shoot breaking under the weight, while others shed snow with little or no damage. This again is a function of both flexibility of the branches and slipperiness of the foliage.

↑ **Two stunted yet vigorous beech trees** growing on a very exposed ridge in the French Alps, with branches and leaves streaming in the gale-force winds.

← **Heavy snow loads the branches** of these Norway Spruce trees. When the weight becomes too great the extremely flexible branches bend down and the snow slips off.

→ **A tornado bears down** on a patch of woodland in central USA. The trees directly in the tornado's path will be destroyed.

↑ **A gnarled old pine** has been killed by lightning strike, its trunk shattered when the water in its wood was explosively converted to steam.

LIGHTNING AND TORNADOES

Lightning strike is a common occurrence in many regions, and trees are the most frequent victims. Usually the tree is killed outright, because the huge electrical current it experiences heats the sapstream; if converted to steam the result is drastic—either the bark is blown off (sometimes down one strip only) or the whole trunk shatters. It is nearly always single trees that suffer, not a whole forest, though sometimes lightning starts forest fires. Another destructive force is the tornado, but it should be noted that tornadoes occur mainly in regions of few trees. And it is likely that, because the total destruction of lightning and tornadoes is visited randomly and on relatively few trees, the tree species exposed to them have not evolved ways to evade or withstand such damage.

Fire and flood

It is not always appreciated just how many of the world's trees and forests are adapted in one way or another to surviving fire. Trees have two main strategies for coping with fire: either they insulate their inner tissues from the heat, allowing the tree to sprout again when the fire has passed; or they depend on regeneration from seed, with fire-resistant fruits releasing seed soon after the fire, or fire-resistant seeds whose germination may be hastened by scorching—or both. Some trees depend on sprouting from buried rootstocks, sacrificing their aboveground stems to the fire. Many larger genera of trees, such as *Pinus*, *Quercus* and *Eucalyptus*, include species that are fire-sensitive (killed by fire) as well as fire-resistant species that can survive most fires. A common feature of the fire-resistant species is the ability to sprout from dormant buds buried deep below thick bark. Trees that depend on seed to preserve their species through fire have a variety of strategies. Many legumes, for example *Acacia* species, shed seeds annually; the hard-coated seeds remain dormant in the leaf litter until scorched by fire. If rain follows, water penetrates the seed coat and germination commences. Many other seeds have their dormancy broken by chemical compounds in woodsmoke.

TREES AND FLOODING

Floods have two kinds of consequence on trees: first mechanical damage to trunks, branches and foliage by strong currents and water-borne debris, also to roots from soil scouring and undercutting; and second, reduction of oxygen supply to the roots as water replaces air in the soil during periods of inundation. Trees that grow on banks of fast-flowing rivers have evolved defenses against damage from strong flood flows, in the form of narrow, tough leaves on flexible branches which can stream in the current. And many other, usually larger, tree species that occur on floodplains have root systems that can survive reduced oxygen levels for a week, or sometimes far longer, when the plain is inundated.

← **This savanna vegetation** in tropical Australia is rich in fire-adapted plant species. Fires are of common occurrence in the tropical dry season and are traditionally used as a management tool by Australian Aborigines in this region, burning small patches at a time which then prevent other wildfires from spreading too far ("mosaic burning"). The most fuel is provided by dead tops of grasses, which have grown tall in the wet season.

→ **Inundated woodlands** on the floodplain of the Mississippi River near Vicksburg, Mississippi.

← **In the aftermath of a wildfire,** this *Melaleuca* tree soon sprouts new foliage from buds deeply buried under the bark of its trunk and lower limbs. The fire was so hot that the smaller branches are all dead. The new green shoots revert to juvenile foliage, and it will be several years before they mature sufficiently to flower and fruit.

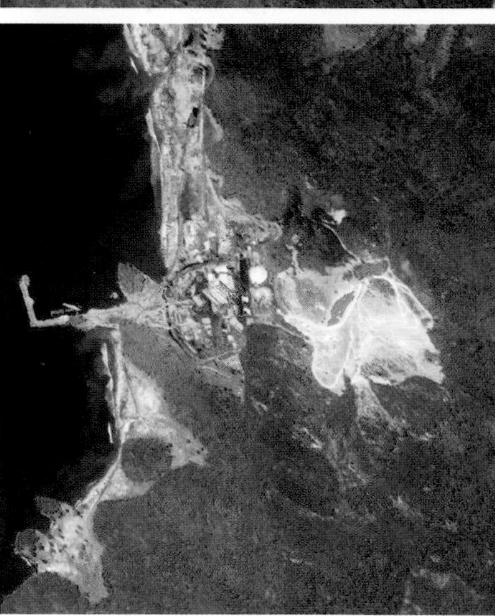

↑ **The above Ikonos satellite images** of an industrial port south of Banda Aceh in Sumatra, Indonesia, illustrate the devastating damage caused by the December 2004 tsunami in the Indian Ocean, the result of an undersea earthquake. The images focus on an industrial plant and pier, as well as an area of forest. The top image shows the area before the tsunami, while the lower image shows the same scene but after the tsunami. As can be seen, a huge stretch of the forest and vegetation was lost along the coast.

Pioneers

Many of the world's most successful tree species are pioneers, which means they are able to rapidly colonize newly created territory, usually in large numbers. The new territory can be of many kinds: coastal or inland dunes piled up by prevailing winds; river sand and gravel bars; estuarine mudflats; moraines of retreating glaciers; major landslips and avalanches; fire-devastated hills; lava, ash and mudflows from volcanic eruptions; and many human-made environments, for example urban rubble left by demolitions. The first requirement for most pioneering plants is an effective means of seed dispersal, preferably over distance and in large numbers. Good examples are many trees of far northern forests, such as birch, alder, aspen and spruce, which have small and abundant seeds that strong winds can carry long distances, effectively blanketing newly created habitats. Some other pioneer trees drop their seeds or pods in water, to be stranded as it recedes, for example the Honey Locust (*Gleditsia*), or many mangroves. In the case of forest destroyed by fire, some successful pioneers are those that leave behind a large store of seeds in the soil, ready to germinate and make rapid early growth, out-competing other trees. Another common attribute of pioneers is the ability to establish in raw mineral soil or sand, and to withstand exposure to harsh conditions such as salt spray off the ocean.

TROPICAL PIONEERS

In tropical regions some of the most common newly-exposed habitats are landslips, caused by heavy rainfall on steep terrain, mudflats deposited by major rivers, and volcanic deposits of various kinds such as the lava-fields in Hawaii and ash beds in Indonesia and the Philippines. Landslips, mostly of limited width, may receive a large supply of seeds from adjacent intact forest, and it is those species that can germinate rapidly and establish in the raw soil and rock that are the pioneers. These mostly make very fast early growth, aided greatly by increased availability of light, but are often short-lived, giving way to slower-growing but long-lived trees as forest gradually re-establishes on the landslip. Volcanic wastes, by contrast, are often very extensive and sometimes isolated, and it is species with very effective distance dispersal that must be the colonizers. Foremost are the ferns, with fine spores that winds carry for tens or hundreds of miles. Following the cataclysmic eruption of Krakatau Island off Sumatra in 1883, ferns appeared in abundance on the bare volcanic ash within a decade or two, and forests of tree ferns developed in parts. Most other trees that subsequently colonized the island had either very light wind-dispersed seeds, or water-carried seeds.

↗ **A seedling of Jack Pine** (*Pinus banksiana*) growing from the ash bed after a forest fire in Minnesota. This pine, native to much of Canada and the Great Lakes region of the USA, has cones that may remain closed for years. Fire will cause release of large quantities of the light, winged seeds, some of which land on ash beds that have already cooled, the soil conditions favoring germination.

← **Regeneration forest** near the peak of Mount St. Helens, Washington. Following the massive volcano eruption in 1980, pioneer species of plants, such as lupines and grasses, began to colonize the newly cleared area on Mount St. Helens. Eventually, pioneer conifers took over, shading out the lower-growing plants below.

→ **The Mountain Ash or Rowan Tree** (*Sorbus aucuparia*) is a fast-growing, but short-lived, pioneer species that occurs widely throughout Europe, extending to the most northerly limits of tree growth in latitude 70° north. The tree can be found growing in mountainous and inaccessible locations.

Diversity and design

Diversity and design

Trees may be divided into three groups according to their evolutionary advancement: the pteridophytes, which include tree ferns; gymnosperms, which include Ginkgo, cycads and conifers; and the angiosperms or flowering plants, which are subdivided into monocotyledons and dicotyledons. Various cacti, succulents and aquatic plants also fall into the tree category.

Types of trees 92

Evergreen and deciduous 94

Ferns 96

Ginkgo 98

Cycads 100

Conifers 102

Flowering trees 108

Types of trees

We can apply broad classifications to the world's trees from a number of different perspectives: for example, geography, where we might start with American trees, Eurasian trees, African trees and Australasian trees. But this is not always useful, since some tree species may fall into more than one group. Or we can divide trees into evergreen and deciduous, but that also has limited value. Classification by climate is perhaps more useful, since it helps to know whether a tree is tropical, warm-temperate or cool-temperate in its growing preferences. But what biologists regard as the most all-purpose classification is what we call a natural classification, which nowadays is the same as saying a classification in which organisms are grouped on the basis of common descent from an evolutionary ancestor. Natural classifications have powerful predictive value: for example, if we know that an obscure tropical tree belongs to the large family Apocynaceae (the dogbane family), then it is a fair bet that it will have milky sap and contain poisonous alkaloids. The most fundamental natural groupings that include trees are quite easily recognized. In order of evolutionary advancement they are the ferns, the ginkgoes, the cycads, the conifers, and the flowering plants or angiosperms, the last subdivided into the primitive or basal angiosperms, monocotyledons and dicotyledons.

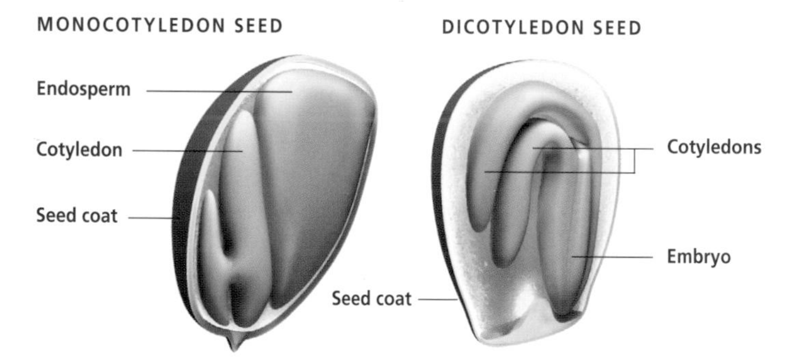

MAJOR TREE GROUPINGS

There are great variations among the major groupings of trees regarding their present-day numbers and the diversity of form, size and organ structure. At one extreme is the Ginkgo, the sole living species of the division Ginkgophyta. Then there are the cycads, with only about 250 living species and diversity within a rather narrow range. The tree ferns number only 5–600 of the known species of ferns, which in total number maybe 10,000. The conifers are hardly more numerous in terms of species than the tree ferns but display much greater diversity in form and size and have contributed to the world's tree cover in quantities out of all proportion to their diversity. But it is the flowering plants, increasingly covering the world's habitable lands over about the last 100 million years, that have achieved levels of diversity beyond the ability of any one biologist ever to comprehend. There are somewhere in excess of 300,000 flowering plant species, many of which reach tree size.

↑ **The gymnosperms** are plants that reproduce by means of an exposed seed ("gymnosperm" literally means "naked seed"). Conifers such as larches (above), cycads and Ginkgo are trees that all fall into this group.

↑ **The angiosperms** are plants that reproduce by means of a seed that is enclosed in an ovary. There are three types of angiosperms: primitive angiosperms, monocotyledons (like the Joshua Tree above), and dicotyledons.

↑ **Most tree species of the world** are angiosperms, and within this group most angiosperm species are dicotyledons. Dicots have a vast range of flower forms, from the classic blossom of apple trees (above) to the blooms of exotic tropicals.

Evergreen and deciduous

From the perspective of a cool climate, such as the USA's New England or Great Lakes regions, or central Europe, the distinction between evergreen and deciduous trees translates almost to the distinction between conifers and broadleaves (flowering plants). This is almost true, but not quite, since even in these climates there are a few evergreen broadleaves, such as hollies (*Ilex*), and a few deciduous conifers, principally larches (*Larix*). However, in warmer parts of the world the picture changes. Here the majority of broadleaf trees are evergreen, especially in tropical rain forests and most southern hemisphere forests. It is true that there are extensive deciduous forests in the tropics, but in those areas deciduousness is a response to the tropical dry season, not to winter frost. Evergreen foliage is the "primitive" or unspecialized state among trees and allows photosynthesis, and therefore growth, the whole year round, obviously an advantage to the tree where climate is favorable. But in climates where there is a season too cold or too dry for growth to be sustained, trees have evolved ways of shutting down to a dormant state where photosynthesis virtually ceases and moisture loss from their tissues is minimized.

CHARACTERISTICS OF DECIDUOUS LEAVES

There are two obvious features of deciduous leaves. First, the leaves are shed cleanly from a predetermined point. This "abscission zone" is a layer across the base of the stalk in which some cells weaken, allowing the leaf and stalk to break off, while other cells below them develop corky walls that effectively seal off the "wound." Second, the leaves lose their green color. In fall, day length and temperature changes trigger a breakdown of leaf chlorophyll, unmasking underlying pigments known as xanthophylls and carotenoids, which give the yellows and oranges; the pigment anthocyanin also develops, giving red or purple.

↓ **Leaves on the Monkey Puzzle** (*Araucaria araucana*) may remain on the tree for 30 years.

← **Shedding leaves, or deciduousness,** is the most common means by which trees become dormant. The leaves are the main organ through which water is lost to the atmosphere—by the process of transpiration, without which photosynthesis cannot take place.

↙ **Unlike the leaves of evergreens,** the leaves of deciduous trees, such as the Silver Maple (*Acer saccharium*), only live for a few months.

↓ **The green glossy leaves of the Blackjack Oak** (*Quercus marilandica*) turn brown in the fall, but often remain attached to the twigs throughout the early winter months. Just before the leaves are shed, the tree draws all the remaining nutrients from them.

Ferns

Although ferns have an ancient lineage, their diverse present-day genera and species are likely to have evolved recently. Ferns are in some sense more "primitive" than flowering plants or gymnosperms, in that their lineage never evolved the capacity to produce wood or seeds. Their ability to disperse remains dependent on minute spores that germinate to produce delicate gametophyte plantlets. The plantlets need permanently moist, humid environments to sexually reproduce and create the next generation of spore-bearing adults. Most ferns are not trees in any sense. But there are two or three fern families containing species that develop aboveground stems of varying heights. Such plants are called "tree ferns," and in most cases their maximum trunk height may be under 10 feet (3 m). However, there are tree ferns that grow tall, as high as 60 feet (18 m) in some cases. The largest genus is *Cyathea*, with around 500 species including all the tallest tree ferns. Its family, Cyatheaceae, is traceable as far back as the Jurassic Period. Dicksoniaceae is the only other family with many trees, and it also goes back to the Jurassic. Its present-day tree fern genera, *Dicksonia* and *Cibotium*, are abundant in some regions.

↓ **The vascular structure of a fern stem.** The outer covering of the stem is quite scaly and provides a strong, armor-like coating. The frond (or megaphyll) contains the spore-producing structures.

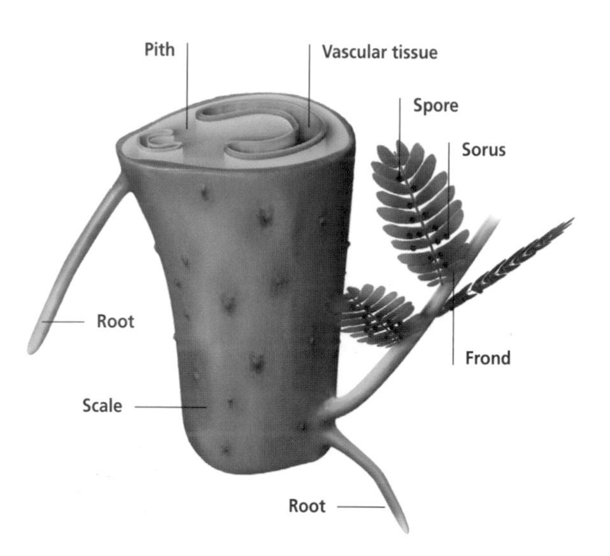

TREE FERN LOCATIONS

Tree ferns are found mainly in the tropics, but are abundant only in places with year-round rainfall and high humidity, such as the mountains of New Guinea, Central America, or the Hawaiian Islands. In the southern hemisphere they extend farthest into the temperate zones, in New Zealand, Tasmania and Chile. In places they form almost pure stands in sheltered ravines or continuous understories beneath taller forest trees. Nearly all tree ferns are single-stemmed, with a single, terminal bud from which large, divided fronds emerge. Each frond is tightly coiled in the bud and rapidly uncurls as the frond expands. Growth is continuous and the bud is never completely dormant. Trunk thickness is constant and does not expand with age.

← **The Australian Soft Tree Fern** or Man Fern (*Dicksonia antarctica*) occurs in large numbers in Tasmania and southeastern Australia. Seldom exceeding 20 feet (10 m) in height, its aerial roots clothe the stem to a diameter of up to 5 feet (1.5 m).

↓ **The underside of the frond** of the Hawaiian tree fern (*Cibotium glaucum*) shows the rows of marginal sori (spore patches) that characterize the family Dicksoniaceae. Each sorus can enclose hundreds of tiny sporangia (spore sacs) that can release thousands of dustlike spores into the breeze.

← **The New Zealand Soft Tree Fern** (*Cyathea smithii*) occurs almost throughout New Zealand and as far south as the offshore Auckland Islands.

↞ **A new tree fern frond** in the "crozier" stage, uncoils from the fern's apical bud. Dense hairs and scales protect the bud from damage.

→ **The black bands of dense fiber** in this trunk cross-section of the New Zealand Mamaku or Black Tree Fern (*Cyathea medullaris*), function like the steel rods in a reinforced concrete pillar.

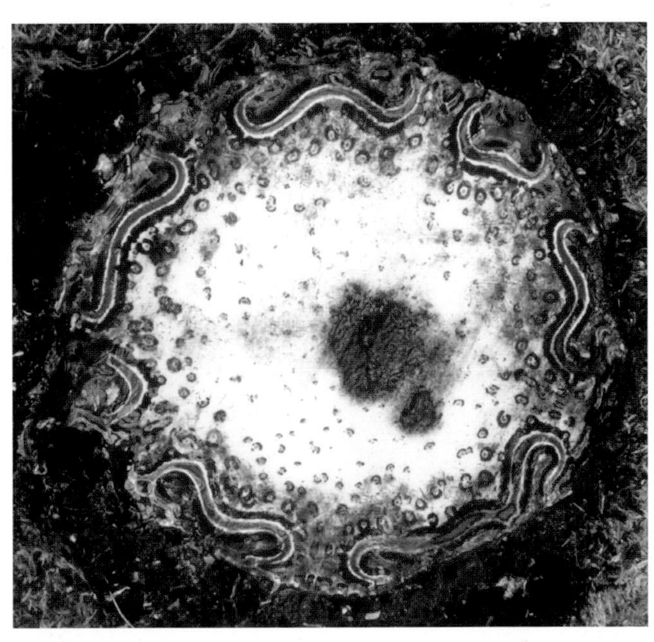

Ginkgo

The genus *Ginkgo* with its single species, *Ginkgo biloba,* is the only living representative of a once more diverse group of trees. It is traceable in fossil records as far back as the early Jurassic period, and the species, with some leaf variation, to the beginning of the Tertiary period around 60 million years ago. During the Cretaceous and Tertiary periods, Ginkgo occurred widely around the world, but now the only living wild trees are found in China, possibly only surviving during the last 1000 or more years because of human cultivation. Fossils related to Ginkgo extend as far back as the early Permian period and it is likely that the tree's lineage diverged from the cycads, conifers, and flowering plant ancestors that existed around that time, almost 300 million years ago. All members of the lineage are termed Ginkgophytes. Ginkgo has a unique mix of characteristics. True wood is produced, increasing steadily in volume for the life of the tree, quite similar to the wood of conifers. The leaves are deciduous, many borne on spur shoots resembling those of cedars (*Cedrus*) or apples (*Malus*). The sexual organs are unusual. The female ovules terminate slender stalks, mostly forked into two at the apex. Pollen, produced from conelike spikes in sacs like flowering-plant stamens, is carried by wind to the ovules where it "germinates" to produce a sperm that swims through a drop of liquid to fuse with the egg cell. The naked seeds are large and fruitlike, with a fleshy, fatty outer coating.

→ **The Ginkgo superficially resembles a dicotyledonous tree,** with its spreading, open-branched habit and broad, deciduous leaves that take on attractive fall colors. But the resemblance does not stand up to close examination. The tree is notable for its adaptability to cultivation and freedom from disease. It is planted in parks and gardens around the world in climates from Finland to Thailand. Extracts of the foliage are claimed to strengthen the human immune system.

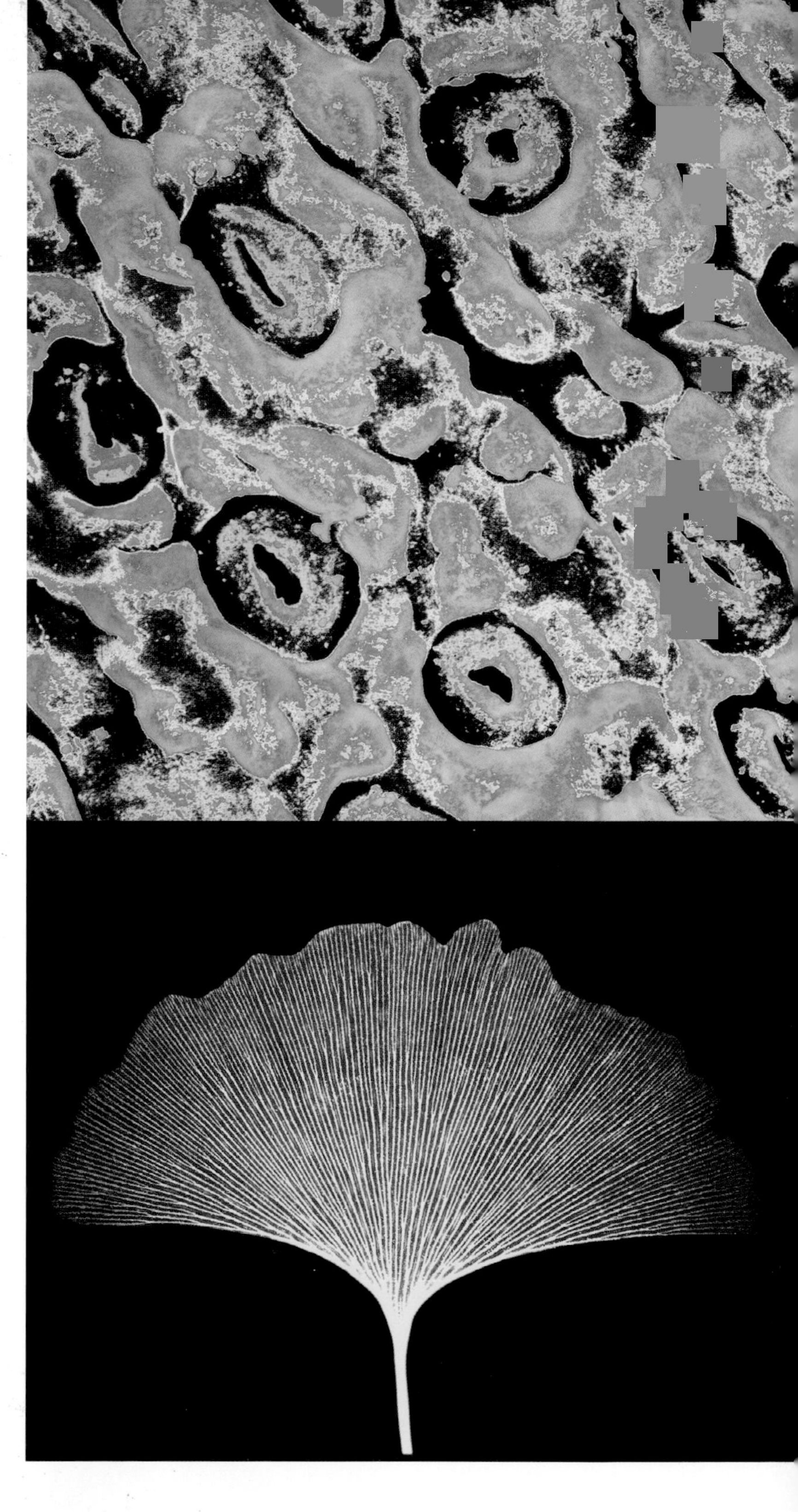

↑ **The "fruits" of the Ginkgo** are actually large naked seeds, one or two on each stalk. They are borne on female trees only, and when fallen the fatty outer layer becomes rancid and foul-smelling. The harder "stone" contains an edible kernel.

↗ **A highly magnified view of the stomata** (the gas-exchanging pores) on the underside of a Ginkgo leaf. The stomata are visible as the merest pale pinpoints to the naked eye.

→ **This X-ray view of a Ginkgo leaf** shows the veins artificially colored yellow. The dichotomous branching of the veins is a feature of some ferns and cycads but never of conifers or flowering plants. Mature Ginkgo leaves such as this example are often not two-lobed.

DERIVATION OF THE BOTANICAL NAME

There is some mystery as to how this tree got the name Ginkgo. It was taken from the alleged Japanese name of the tree or of its seed, as heard by the German naturalist and physician Engelbert Kaempfer during his sojourn in Japan in 1689–92. But he must have heard it wrongly, since "ginkgo" has little resemblance to the actual Japanese name. Nevertheless, it was adopted as the scientific name by Linnaeus and became the common name in the West, familiar to herbal medicine enthusiasts as well as gardeners. The epithet *biloba* refers to the two-lobed leaves, which have given rise to one of its Chinese names, *yajiao* meaning "duck's foot." From its probable last natural stronghold in southeastern China, the Ginkgo was spread by humans to northern China, Korea and Japan well over 1000 years ago, and was introduced to Europe about 1730.

Cycads

Cycads have a popular association with the notion of "living fossils," which is not quite accurate since they have diversified and thrived in various parts of the world in quite recent geological history. But it is true that their ancestral forms in the age of dinosaurs were much more prominent in Earth's vegetation. The cycadophytes are one of the five surviving major lineages derived from the Paleozoic "lignophytes" (plants able to produce wood), their lineages diverging between about 300 and 200 million years ago. Present-day cycads include more shrubs than trees among their 250 or so species, and the largest tree cycads are no more than 60 feet (18 m) tall. There is little evidence that their ancestral forms grew much larger than this. Cycads are typically of a palmlike form, with a thick, unbranched or few-branched trunk topped by a crown of radiating pinnate fronds. The trunk does contain wood of a sort, produced by a cylinder of cambium, but rather little of it and with a large core of starchy pith. An outer layer of corky tissue and old frond bases provides additional support. Reproductive organs are borne in large cones, the male and female always on different plants. Fleshy female cone scales bear ovules on their margins, these developing into large naked seeds when fertilized by pollen from the elongated male cones.

GENERA AND DISTRIBUTION

Cycads are more widely distributed through the tropics and subtropics than has often been suggested, though absent from most arid regions. The living species are divided among three families: Cycadaceae, with the single genus *Cycas* of about 90 species; Stangeriaceae, with two genera *Stangeria* and *Bowenia* of one and two species respectively; and Zamiaceae, which includes all nine remaining genera. The greatest diversity of genera is found in eastern Australia and Central America, each with four genera. In Asia only *Cycas* is found but its species are diverse and quite common, from eastern India to southern Japan. South Africa is home to one of the most curious genera, *Stangeria*, with creeping rhizomes and stiff, fernlike fronds with dichotomous veining. *Bowenia*, its Australian relative from rain forests of northeast Queensland, has a similar habit but with taller, glossy fronds that are twice-divided.

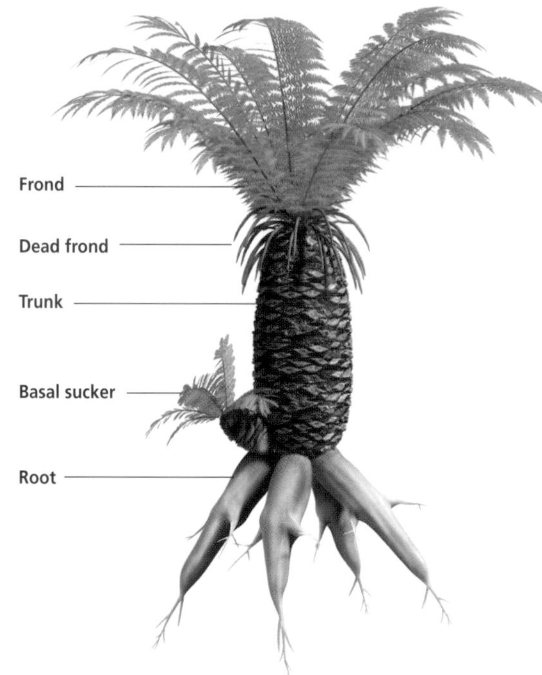

↑ **The structure of a cycad.** The strong, fleshy roots of some cycads are contractile, meaning that they contract and pull the plant further into the ground.

← **A young cycad frond.** The genus *Cycas* is distinct from all other cycads, except *Stangeria* and *Bowenia,* in having fronds whose leaflets uncoil from the bud, rather like the fronds of ferns.

↑ ***Cycas revoluta*** is popularly known as "Sago Palm" but it is neither a palm nor the true source of sago. Now the most widely cultivated cycad in the world, its origin is Japan's southerly Ryukyu Islands, where stands such as this grow on steep, rocky slopes.

→ **A feature of many cycads** is the production of a single flush of new fronds each year from the stem apex. They alternate with whorls of highly modified scale-like leaves, termed cataphylls, that protect the terminal bud.

Conifers

Conifer literally means "cone-bearer" and refers to the way these trees carry their seeds on the scales of cones, rather than enclosed in a fruit developed from a flower, as in the flowering plants. It should be noted that conifers are not the only cone-bearers: the cycads have cones of similar construction. And some conifers, such as yews and podocarps, bear their seeds on organs that are hardly recognizable as cones. All conifers have cones of a second type, the pollen cones, usually smaller than the seed cones, which produce wind-borne pollen from numerous small pollen sacs arranged variously on delicate scales. Conifers, like the cycads and Ginkgo, are gymnosperms, a group now regarded as a stage of evolution rather than one of common descent. Although steadily pushed aside by the flowering plants from about the late Cretaceous period onward, conifers have remained a vigorous and still-evolving component of the world's flora, especially in cold climates. Unlike flowering plants, however, they have never evolved lifeforms such as herbaceous perennials and annuals, bulbs or climbers—though there is one conifer from New Caledonia that has become parasitic. They remain obstinately woody, growing to trees or at least shrubs, and they include the world's tallest trees as well as the longest-lived plants.

CONIFER CHARACTERISTICS

The conifers, or to give them their formal name the division Pinophyta, include 600 species divided among 64 genera, these in turn are divided among seven families. There is considerable diversity of form among them but they have certain characteristics in common, in addition to their reproductive organs. The leaves are often needle-like, or if broader and flat they have veins arranged, not in a network (reticulate), but lying parallel, and the leaf texture is thick and leathery. There are cavities or "canals" filled with aromatic liquid resin running through leaves, bark and wood. And the wood structure is simpler than that of flowering plants, lacking the larger vessels (pores) and with only small, inconspicuous medullary rays, so the wood appears fine-grained.

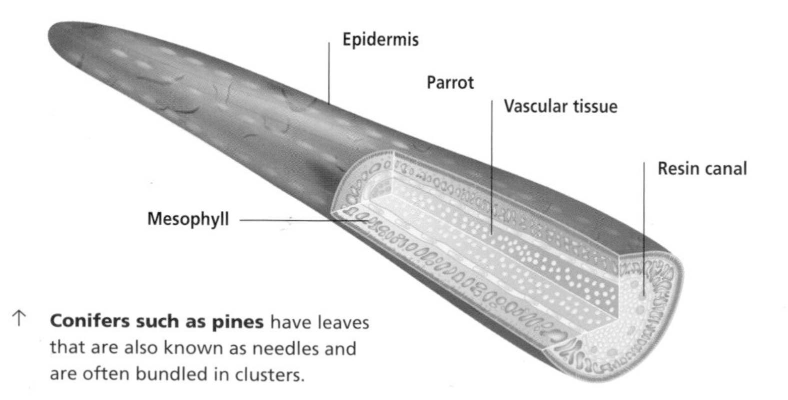

↑ **Conifers such as pines** have leaves that are also known as needles and are often bundled in clusters.

← **A female cone of the Wollemi Pine (*Wollemia nobilis*).** Dscovered only as recently as 1994, this tree has leaves that are arranged in four ranks.

↞ **The deciduous conifer Dawn Redwood (*Metasequoia glyptostroboides*)** has two-rowed leaves on short shoots, the whole shoot sheds in the fall.

↙ **The tallest trees in the world,** Coast Redwoods can grow to heights of more than 350 feet (115 m). They inhabit west-coastal USA, mainly in northern California.

↓ **Junipers are true conifers** from the northern hemisphere, most of which are cold-hardy. Old specimens can become twisted and gnarled in appearance.

Conifers continued

The pine family (Pinaceae) and cypress family (Cupressaceae) include nearly all the conifers dominating temperate northern hemisphere coniferous forests. With 11 genera and about 230 species, the pine family is pre-eminent in the great boreal forests, where its genera *Pinus* (pines), *Larix* (larches) and *Picea* (spruces) occur in countless millions, though with few species of each. Although it barely reaches the equator (in Sumatra), this family has its greatest diversity in the middle latitudes of the northern hemisphere and even into tropical highland regions in Mexico. The true pines (*Pinus*) constitute its largest genus, with over 100 species. These have a highly distinctive structure, with two kinds of scale-leaves as well as needle-like green photosynthetic leaves that are grouped in clusters, the number of needles per cluster almost fixed for each species: two, three, or five in most cases. Each cluster is interpreted as a lateral shoot that has lost any capacity to grow farther.

CROSS-SECTION OF A FEMALE CONE

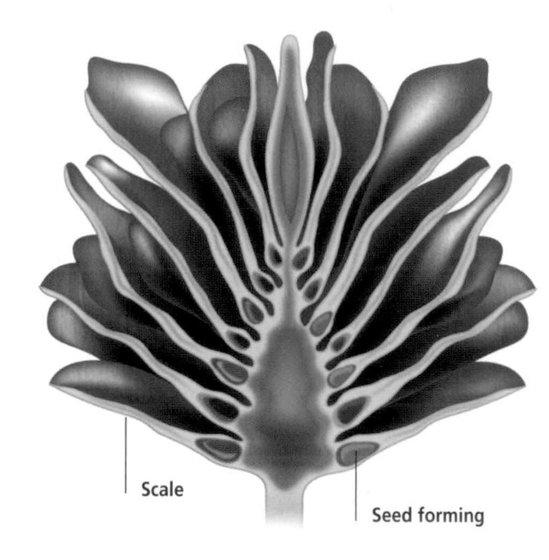

↑ **Cones are seed-bearing structures** that conifers produce instead of flowers. Male and female cones are produced on the same tree. The male cone bears pollen sacs, and this pollen is dispersed by the wind and transferred to the female cone, where fertilization takes place.

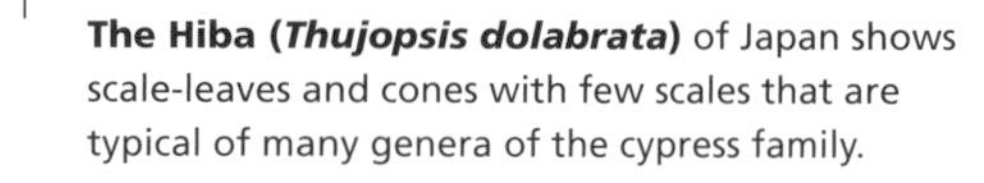
↑ **The Hiba (*Thujopsis dolabrata*)** of Japan shows scale-leaves and cones with few scales that are typical of many genera of the cypress family.

← **A pair of Douglas Firs** (*Pseudotsuga menziesii*) growing in Bryce Canyon National Park in Utah.

CYPRESS FAMILY

The cypress family shows some of the greatest diversity of all conifers, with about 140 species divided among 25 genera distributed through both the northern and southern hemispheres, though many of the genera have only one or two species. Some have been more successful than other conifers in adapting to dry climates, especially *Juniperus* (junipers) and *Cupressus* (cypresses) across the northern hemisphere, and *Callitris* (cypress pines) in Australia. These are all trees of relatively compact size with leaves mostly reduced to tiny scales, closely pressed against the branchlets and often with a waxy cuticle, to reduce water loss.

↗ **A twisted gnarled trunk** of an ancient Bristlecone Pine (*Pinus aristata*) in the Sierra Nevada, California.

→ **The small needle leaves** of Atlas Cedar (*Cedrus atlantica*) are mostly borne on rosette-like short shoots.

→→ **The flat-crowned *Pinus hwangshanensis*** clings to cliffs of the Huangshan World Heritage Site in Anhui Province, China.

Conifers continued

The family Araucariaceae or araucaria family is notable among conifers for its almost exclusively southern hemisphere occurrence. It consists of three genera of stately, often very symmetrical, trees, namely *Araucaria*, *Agathis* and the recently discovered *Wollemia*, containing respectively 19, 21 and one species. While *Agathis* has a regional distribution from New Zealand and other southwest Pacific Islands to the Malay Peninsula, *Araucaria* is more fragmented, with two species in South America, 13 in New Caledonia, one in Norfolk Island, and two each in eastern Australia and New Guinea; and the sole species of *Wollemia* is known from a handful of trees in New South Wales. The family has a fossil record dating back to the Jurassic. The yew family, Taxaceae, by contrast, is found mostly in the northern hemisphere, though one of its five genera is endemic to New Caledonia. All its 17 species have flattened, narrow leaves and seed cones reduced to mere knobs, with a fleshy outgrowth (aril) partly or wholly enclosing each of the one or few seeds. They are trees of modest size, or shrubs.

↑ **An aerial view of taiga, or boreal forest,** in Siberia. Taiga forest is largely inhabited by conifers such as larch, spruce, fir and pine.

↓ **These ancient cedar trees,** which thrive in mountainous conditions, cling resolutely to a rocky ledge in the North Woods, Minnesota.

THE PODOCARPS

Members of the family Podocarpaceae are referred to as the podocarps. This interesting group is most diverse in the southern hemisphere but is also well represented in the wet tropics, while a few species are found as far north as Japan and central China. It was once treated as comprising only three genera, *Podocarpus*, *Dacrydium* and *Phyllocladus*, but the first two of these have been split by botanists into a total of 17 genera, while the bizarre *Phyllocladus* is now thought to merit a family of its own. *Podocarpus* still remains the largest genus with 105 species and occurs throughout the range of the family. It has flattened leaves, quite large in some species, and seed cones modified into a fleshy and even juicy stalk, carrying often just one large seed. The trees, sometimes called "plum-pines" or "plum-yews," are found mostly in rain forests. *Dacrydium*, by contrast, has fine, rather feathery foliage with needle-like leaves. Some of its species are common trees of the mossy forest zone in mountains of Southeast Asia and the Malay Archipelago.

→ **Even huge conifers begin life as tiny seedlings.** Seeds are formed inside the female cones and then drop to the ground where they germinate.

↑ **Spruces (*Picea*) are generally found in the world's northern forests,** such as the Riisitunturi National Park in Lapland Province, Finland.

↓ **The striking yellow pollen cones of the Loblolly Pine** (*Pinus taeda*). This species, native to southeastern USA, prefers a moist habitat.

Flowering trees

Flowering trees are all those trees belonging to the Magnoliophyta or "angiosperms," the group of seed plants that has taken over most of the world's lands during the last 30 million years or so. Angiosperms started from obscure beginnings, the molecular evidence suggesting their ancestors may have split from other seed-plant lineages as far back as the Carboniferous period, around 300 million years ago; although fossils with identifiable flowers have not been found any older than late Jurassic, around 140 million years back. For much of the Cretaceous and earlier Tertiary they seem not to have been a dominant part of the vegetation, though it seems likely that by the mid-Cretaceous most of the present-day families had appeared in some form, indicating a high level of diversity even then. Since then, angiosperms have evolved into by far the most diverse of any of the forms of green plant life, with somewhere between 300,000 and 400,000 species. Much of this diversification has taken place in families that contain no trees or a minority of trees, such as the very large grass, orchid, daisy and legume families. But in the tropics, for example in the Amazon, it can be surprising to find just how large a proportion grow into trees. It is doubtful that any botanist has ever attempted to count the total number of tree species in the world, but it is quite possible that more than 20 percent of all flowering plant species reach tree size.

FLOWERING PLANTS AND INSECTS

The features that most obviously define flowering plants are, first, that the developing seeds are fully enclosed within the fruit tissues; and second, that they produce flowers. There is a strong relationship of mutual adaptation between flowers and the fauna that pollinate them, principally insects. It has often been said that the flowering plants could not have evolved any of their present great diversity without insects and, conversely, that a large part of present-day insect diversity would not have evolved without flowering plants. It is entirely due to the need to attract insects or birds for pollination that flowers have become the showy, often scented and colorful structures we know so well.

↓ **The stunning fall foliage and gnarled trunk** of the Japanese Maple (*Acer palmatum*).

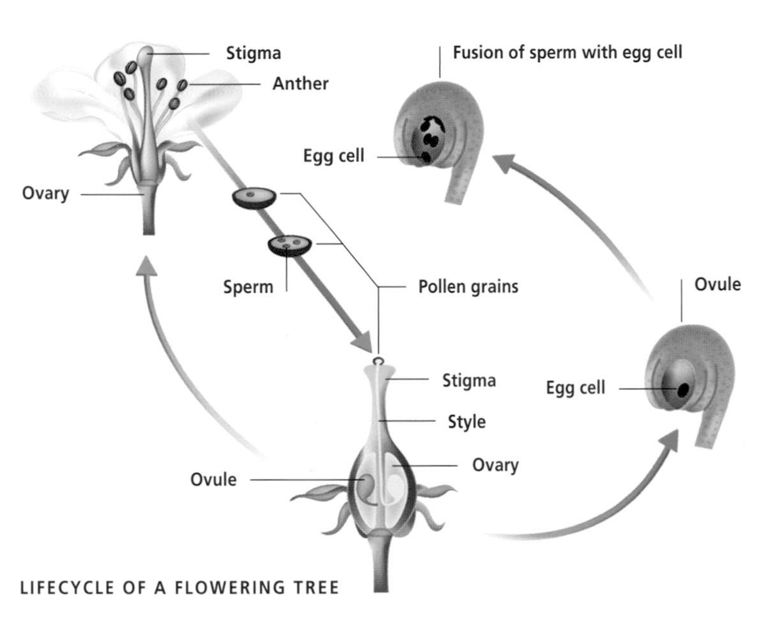

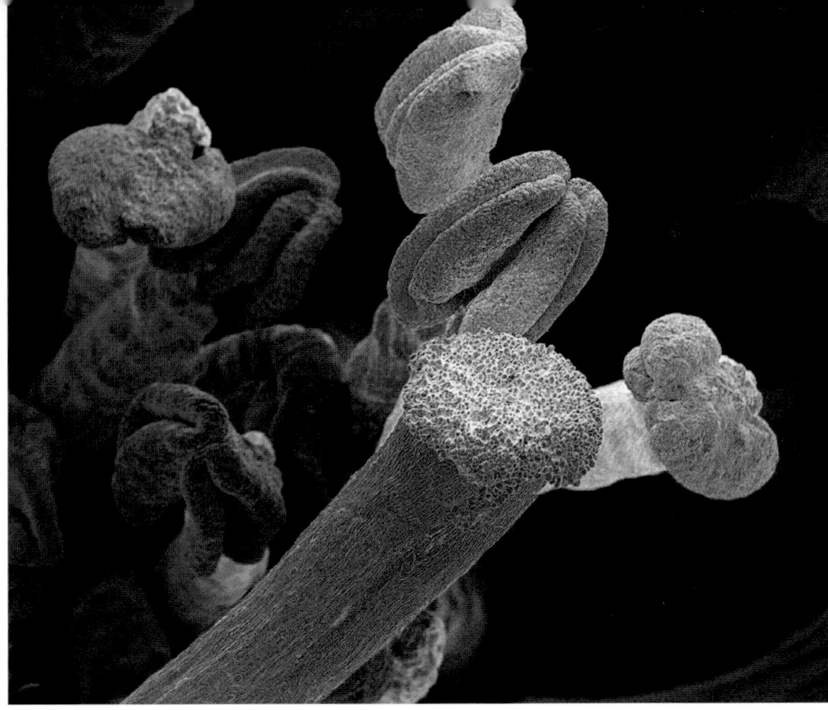

↑ **A peach tree blossom** magnified 100 times, showing the female stigma and style and male anthers.

↖ **The pollen (containing sperm)** is transferred from the male anthers to the female stigma, via the style. The pollen then makes its way down to the ovary where it enters the ovule to fuse with the egg cell.

← **There are approximately 30 species** in the *Tilia* (linden or lime) genus in the northern hemisphere.

↓ **Crabapples** are relatives of the cultivated apple, but have small, sour fruit. Some crabapples are grown in apple orchards to assist with pollination.

Flowering trees continued

For about two centuries, up until almost the year 2000, it was believed by botanists that the flowering plants could neatly be divided into two major groups: the monocotyledons and dicotyledons (monocots and dicots for short), their distinguishing features (specific differences in the flowers, pollen, seeds, stems, roots and leaves) being evident to any school biology student. It came as something of a shock, therefore, when evidence from molecular and other studies began to make it clear that there was a group of plant families that must have split off from the angiosperms' main "stem" before that stem divided into the monocots and dicots. To conform with the taxonomic doctrine that a clade (a branch of an evolutionary "tree") must contain all its members, and only its members, it is necessary to exclude these families from both the monocots and dicots, so they are now referred to by taxonomists as the "basal angiosperms" or sometimes "primitive families." Their best known members are the magnolias, but they comprise at least 25 families and some thousands of species, a large proportion of them trees, which had all previously been called dicots. Most numerous are the laurel family with about 3000 species and the annona family with about 2000, both found mainly in tropical rain forests.

BASAL ANGIOSPERMS
Trees belonging to the basal angiosperm families tend to have flowers without the regular distinction between petals and sepals as found in other families, but with rather less differentiated, fleshy perianth segments, these sometimes numerous. The flowers frequently have fruity or spicy perfumes, strongest in the evening and mostly attracting beetles, moths or bats, which may eat parts of the flowers in exchange for pollinating them. Aromatic oils and hot-tasting compounds occur widely in their bark and leaves.

↙ **The Camphor Laurel (*Cinnamomum camphora*)** is best known for its pungent essential oil.

↓ **The pollen grains** of monocots have one furrow or pore, unlike dicot pollen grains which have three.

← **Belonging to the annona family,** the deciduous North American Pawpaw (*Asimina triloba*) has fetid-smelling flowers with six tepals that appear before the leaves. The tree is best known for its segmented green fruit with sweet yellow flesh.

→ **The flowers of the Bull Bay Magnolia** (*Magnolia grandiflora*) are typical of the basal angiosperm group, with wide flat tepals and many stamens. The tree is a popular ornamental in warm-temperate areas.

↓ **Tulip Tree (*Liriodendron tulipifera*),** native to eastern North America, is a member of the magnolia family. Its flowers have nine tepals and numerous stamens. The deciduous leaves have a unique shape, three-lobed with the wide center lobe squared or shallowly notched. It is an important timber tree.

Flowering trees
continued

The earliest monocots of the Cretaceous period seem to have included tree forms, notably the palms. But from the mid-Tertiary onward the monocots underwent a great burst of evolution that produced the present dominance and diversity of grasses on the world's open plains, and an even greater diversity of orchids. These two largest families, together with other large monocot families such as sedges, rushes and lilies, consist almost entirely of herbaceous plants, and that is the direction in which the monocots have chiefly evolved. Even the palms, which we think of as treelike, have at least half their 2500 or so species with a climbing or shrubby growth form, and the same goes for the pandans, yuccas, dracaenas and other monocot groups that contain some trees. It could be argued that the more massive monocots are evolutionary relics, but we cannot be sure that some of them have not recently evolved from lower-growing forms. That still leaves numerous palms, and in some regions pandans, as well as significant trees of many landscapes, especially in the tropics.

MONOCOTS AND DICOTS

Monocot trees differ fundamentally in structure from dicot trees. Their leaves have closely parallel veining and are attached to the stems by encircling bases. Their stems grow from massive terminal buds that lay down most of the tissues that the tree will need; in the case of single-trunked palms, if that bud is cut off the palm will die. And their conducting tissues are in the form of numerous "vascular bundles" scattered through a softer ground tissue, each bundle containing strong fiber cells as well as xylem and phloem. Although dracaenas and yuccas (but not palms and pandans) develop a cambium layer encircling their trunks, this cambium lays down discrete vascular bundles on its inner side rather than wood as in conifers, dicots and basal angiosperms. Tree monocots seem more limited in size than dicots, their ultimate height generally under 50 feet (15 m) except for a few palms that can exceed 100 feet (30 m); and one palm from the Colombian Andes, *Ceroxylon*, is reported to grow as tall as 200 feet (61 m).

↑ **Coconut Palms (*Cocos nucifera*)** on the beach in Punta Cana, Dominican Republic. The thickening of the trunk base in such large palms results from opening-up of air spaces between the cells, not from a cambium layer.

← **Fruiting branches of the Date Palm (*Phoenix dactylifera*)** with green immature fruit. This palm grows best in hot, dry climates but requires constant soil moisture.

↞ **A succulent evergreen swordleaf,** the Quiver Tree (*Aloe dichotoma*) has a stout trunk that thickens with age by growth of a cambium layer.

↓ **The Dragon Tree (*Dracaena cinnabari*)** is unique to the Socotran Archipelago off the coast of Somalia. The flowers of this slow-growing but long-lived tree may not appear for 20 years. The lateral branches appear after flowering.

Flowering trees continued

Flowering plants have developed countless forms in the course of their long evolution. Some have obvious elaborations but others are more mystifying, such as the tendency toward compound leaves, flowers and fruits that are found so frequently among flowering plants. Why, for example, do mulberries (genus *Morus*) have a dense spike of small female flowers that develops into an apparent single berry, quite like the fruit of blackberries (genus *Rubus*) that in fact develop from a single larger flower? Consider also the spiky compound "fruits" of the north-temperate *Liquidambar* and the Australian *Casuarina*—formed in the first from fused sepals and capsules, in the second from fused floral bracts, both releasing small winged seeds but in the case of *Casuarina* the "seeds" are in fact small, dry fruits. What these examples illustrate is one of the more pervasive features of flowering-plant evolution, namely convergence—the tendency for unlike structures in unrelated plants to be pushed by natural selection in the same direction, resulting in new structures that appear identically adapted. But what is still so often a mystery to us is just why the adaptation should have taken that form.

↓ **The Goat Willow (*Salix caprea*)** is native to Europe and northwestern Africa and gets its name from the wild goats which enjoy the foliage. The tree is also known by another common name, Pussy Willow, because of its downy catkins which somewhat resemble cat's paws.

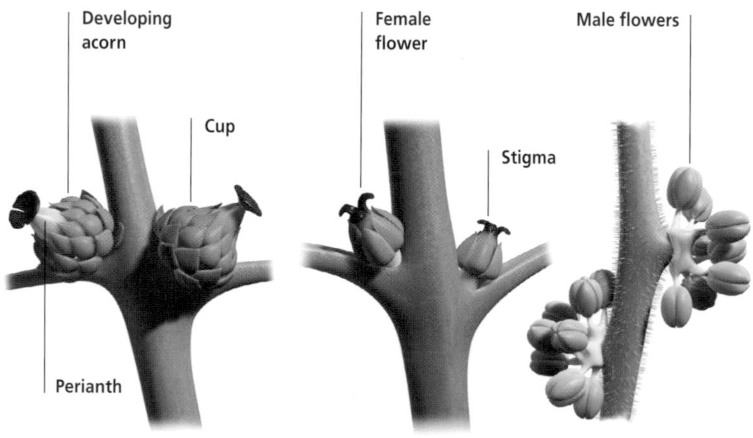

↑ **Sessile Oak (*Quercus petraea*) fruits and flowers.** Although the Sessile Oak is monoecious (male and female flowers on the one tree), the male and female flowers are borne on separate inflorescences. The acorns, which develop in fall, are the fruit and contain a single seed.

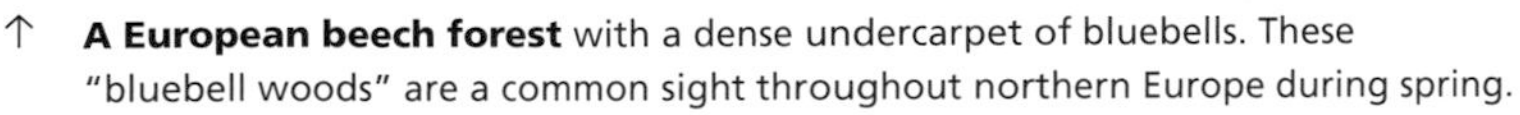

↑ **A European beech forest** with a dense undercarpet of bluebells. These "bluebell woods" are a common sight throughout northern Europe during spring.

↗ **The white trunks of Quaking Aspens (*Populus tremuloides*).** Due to the flattened petioles, even the slightest breeze sets the leaves of this tree fluttering or "quaking," hence its common name.

→ **The ornamental cherry** is native to the Himalayas and eastern Asia. The Yaezakura type of cherry blossom has large, double-petaled pink flowers.

HUMAN USES OF FLOWERING TREEES

Humans, and especially humans as cultivator, have been very much latecomers onto the scene of flowering-plant evolution, but have taken endless advantage of the products of these plants' hundred-million-plus years of diversification. When it comes to flowering trees, we think first of fruit, and it is really the chemical diversity of fleshy fruits, derived from so many plant families and genera, that has so enriched human food culture. Think of the difference, for example, between an avocado and an orange in aroma, flavor and texture. In a quite different direction, consider the diversity among timbers of flowering trees, so much greater than that of conifers. This great diversity has been much exploited by humans though not (unlike that of crop plants) influenced by them, having evolved many millions of years earlier.

Flowering trees continued

In warmer climates, the diversity of flowering trees is expressed to its fullest degree. Trees, and indeed woody lianes and shrubs—since there is no clear distinction between these and trees—have evolved countless ways of coping with different types of physical environment such as climate, soils, sea coasts, mountain crags or deep ravines; and of their biotic environments, which encompass every kind of interaction with other plants, animals and fungi, including competition with their tree neighbors. Adaptations to all these environmental factors are found in each and every tree species. Many of these adaptations are covered in other parts of this book, but what is so remarkable about flowering trees is the almost unbelievable diversity of the flowers and fruits. Once animals (including birds and insects) became the major agents of pollination and seed dispersal, probably in the late Cretaceous period, evolution drove both plants and fauna into such a multitude of co-dependent relationships that biologists even now are only just beginning to appreciate their extent.

TROPICAL TREE FERTILIZATION

Genetic variation in a species is ensured by the transfer of pollen between different trees of the species (outbreeding), so flowers have many mechanisms to encourage this. The showy trumpet flowers of the tropical *Jacaranda* and *Tabebuia* require specific insect and bird pollinators, while the pendulous, fruity-scented flowers of trees like *Adansonia* are aimed at pollination by nocturnal bats. Other narrow-tubed flowers smelling sweetly in the evenings are pollinated by moths, often with a high specificity between tree and moth species. And the sole purpose of a fruit, likewise, is to provide a means of seed dispersal. In the tropics it is likely that a majority of tree species, including all the fleshy-fruited ones, depend at least partially on animals for seed dispersal, and of course there are numerous kinds of fauna that depend on tree fruits for food and in doing so disperse the seed.

↑ **The germination process of a beech tree,** from the emergence of the embryonic root, through the expansion of the seed leaves, to flowering.

↑ **An ant collecting** protein-lipid Beltian bodies from the leaflet tips of a Swollen Thorn Acacia, to feed to its larvae. The only known function of the Beltian bodies is to provide these ants with food, and in turn, the stinging ants guard the plant from predators.

← **A mature beech tree** in fall, with its leaves colored orange and ready to be shed. There are 10 species of beech (*Fagus*), and all are native to the temperate regions of North America, Asia and Europe.

↗ **A view looking up inside the interlocked branches** of a Strangler Fig (*Ficus benjamina*), which has successfully taken over and killed its host tree, of which there is nothing left.

→ **The unusual spiky fruits of the Rambutan (*Nephelium lappaceum*)** are particularly popular in Southeast Asia, where they are widely cultivated. The fruit has a leathery red skin which is covered in greenish, fleshy thin spines.

Flowering trees continued

Perhaps the most significant thing that can be said about the species of dicot trees that dominate temperate northern hemisphere forests is that they all show strong relationships with trees from closer to the equator, mostly in rain forests or subtropical hill forests. Nearly all the well known genera of deciduous trees have species in tropical forests, some such as *Acer* (the maples) with a small minority of tropical species, but others such as *Quercus* (oaks) with almost a majority. There is a common pattern of the tropical representatives, or at least some of them, being evergreen while their temperate relatives are deciduous. One of the standard explanations of this is that the opening up of vast new habitats following periodic icecap retreats between the late Tertiary and Quaternary glaciations gave opportunities for some tree genera in the subtropics to rapidly colonize those habitats and evolve many new forms better adapted to cool-temperate climates. One major adaptation was deciduousness, another was pollination by wind, maybe due to lack of insects in those treeless wastes. Wind-pollinated plants do not need showy or scented flowers but do need large quantities of stamens and a matching large exposure of stigmas, hence the high frequency of the catkin form of inflorescence in many northern deciduous trees. The most abundant of these trees belong to a quite small number of genera compared with trees of tropical and subtropical forests.

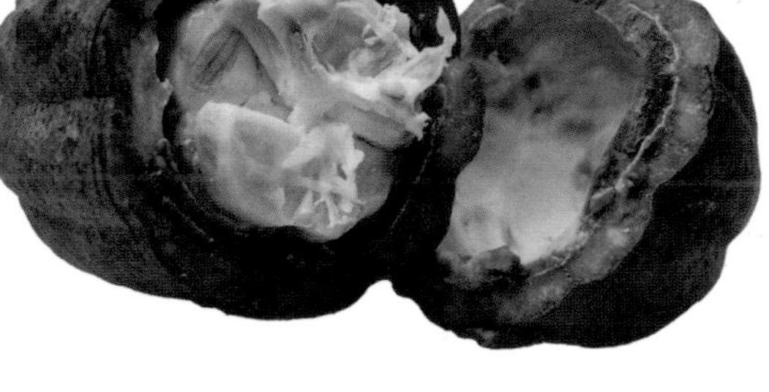

WIND POLLINATION AND SEED DISPERSAL
Salix (willows) and *Populus* (poplars) of the family Salicaceae have highly specialized catkins, male and female always on different trees, and downy seeds that float long distances in the lightest breeze. Salix especially has evolved numerous species, even in the far north. *Ulmus* (elms) shows parallels to both these, though its small wind-pollinated flowers are hardly catkin-like and male and female are on the same tree; and its seeds are sealed in papery fruits carried quite far by strong winds. *Acer* (maples) shows similar pollination and dispersal strategies to those of *Fraxinus*. What is interesting is that these groups of northern deciduous trees come from a range of quite unrelated plant families, all far more diverse in the tropics, and the temperate tree genera all show close links to tropical genera of the same family.

← **Frangipani (*Plumeria rubra*)** flowers have a slight fragrance in the evenings, attracting nocturnal pollinating insects. Although the flowers have no nectar, moths are attracted to the fragrant flowers and unwittingly transfer the pollen from flower to flower while they search for the non-existent nectar. Like its relative, the poisonous shrub oleander, the Frangipani has milky sap in its stems and leaves.

↞ **A cacao fruit** from the Cacao tree (*Theobroma cacao*) has been broken open. Its beans are embedded in the edible, fatty flesh.

← **Showy flowers,** such as the South American *Tibouchina* genus, are brightly colored in order to attract pollinating insects.

→ **Snow Gums** (*Eucalyptus pauciflora*), with their characteristic smooth white trunks and twisted branches, are found in subalpine areas of eastern Australia, such as the Snowy Mountains.

Palms

The palms, which constitute the family Arecaceae, include all the largest monocot trees. But the image of a crown of long fronds at the top of a single tall trunk, as typified by a coconut palm, gives a very incomplete idea of the diversity of this mainly tropical family of approximately 190 genera and 2500 species. The tall tree palms show great variation in form: many are multi-stemmed, branched from ground level, and there are a few, such as African *Hyphaene* or doum-palm species, that have trunks repeatedly forking high above the ground—some of the few cases of truly dichotomous branching among flowering plants. There are also many tropical palms with very short trunks but massive upright fronds that in the African *Raphia* may reach lengths of up to 80 feet (24 m). What we call the "frond" of a palm is in fact a single leaf, though mostly divided into many segments or leaflets, so these larger palm fronds are truly the largest leaves of any plants. Palms form an isolated lineage that goes back to around 120 million years in the early Cretaceous. Their stems, sometimes exceeding 100 feet (30 m) in height, contain no wood and have no capacity to increase in diameter except by the opening up of air spaces between the cells; their strength derives entirely from their strongly fibrous vascular bundles and internal sap pressure keeping other cells rigid.

PALM BY-PRODUCTS

Palms include many species of importance to humans and a few of worldwide commercial value, principally the Coconut, the Date and the African Oil Palm. With few exceptions their fruits and oil-rich seeds are edible, if not to humans then to many other types of fauna. Coconuts and oil palms have seed oils useful for both edible and non-edible purposes, and the Oil Palm has an additional fruit oil with properties different from the seed oil. South America in particular has many other palms yielding useful oils, as well as species in the genus *Euterpe* that yield a vegetable (palm hearts) from the terminal bud or a popular drink (açaí) from the fruit flesh.

↑ **Palm trees** growing in a flooded region of the Corrientes province near Goya, Argentina.

← **An orchard of Date Palms (*Phoenix dactylifera*)** in southern California. The ripening fruit have been protected from rain by paper cones

↗ **The large seed of the Coconut Palm (*Cocos nucifera*)** is often carried by ocean currents to new germination sites. The seeds can remain viable for up to 2 weeks in the ocean.

↠ **The distinctive stilt roots of the tree palm *Iriartea deltoidea*,** native to Central and South America.

→ **A cross-section** of a palm trunk reveals the hundreds of vascular bundles that carry water to the leaves.

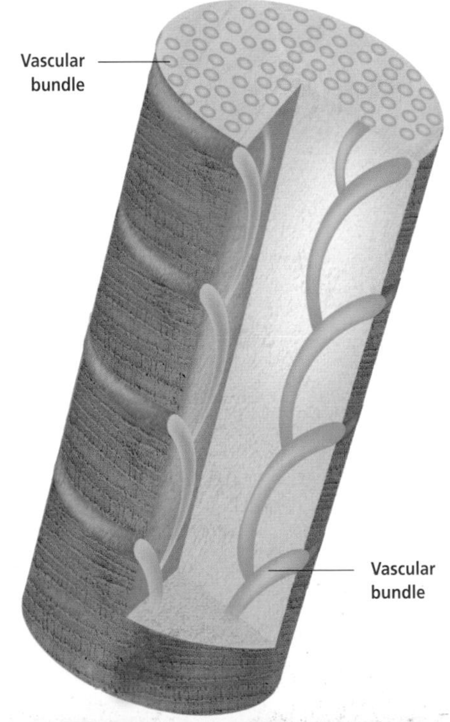

Cacti and other succulents

There is an awkward relationship between the concepts "cacti," "succulents" and "trees." Cacti (plural of cactus) are members of the plant family Cactaceae, native to the Americas, nearly all of which are stem succulents though a minority have fleshy leaves. The name was adopted by Linnaeus who placed many of the species in the one genus *Cactus*, though the Greek word he used referred to unrelated plants. Succulents are plants with fleshy, water-storing tissues in either stems, leaves or both, and they include representatives from a range of plant families including the Cactaceae. Many are adapted to arid climates with fiercely hot sun and dry air; nearly all have a physiological mechanism known as CAM (Crassulacean Acid Metabolism) whereby their stomates open only during the cooler, more humid nights and the carbon-dioxide admitted is stored in plant acids to be released for photosynthesis later in the daytime. And of course to be regarded as a tree, a plant should fit the criteria discussed on page 18.

TREELIKE SUCCULENTS

A relatively small proportion of succulent families and genera include species that develop treelike proportions. In the Americas, the Cactaceae stand out. These are mostly the "cereoid" genera, which include the Saguaro and Organ-pipe Cactus of the USA, however the great majority are found in Mexico, the Caribbean or South America. Africa, the other continent rich in succulent plants, has more diverse tree succulents. The most prominent are the tree euphorbias and the tree aloes. The large genus *Euphorbia* has species that grow as tall as 60 feet (18 m). Most tree forms occur in southern and eastern Africa but some spill over to Arabia and even tropical Asia. A similar pattern is found in the mainly African genus *Aloe*, though its tree species do not get beyond southern Arabia. There are miscellaneous other genera from families that contain pachycaul (swollen-trunked) trees, which in a sense are stem succulents. They include plants from all the continents.

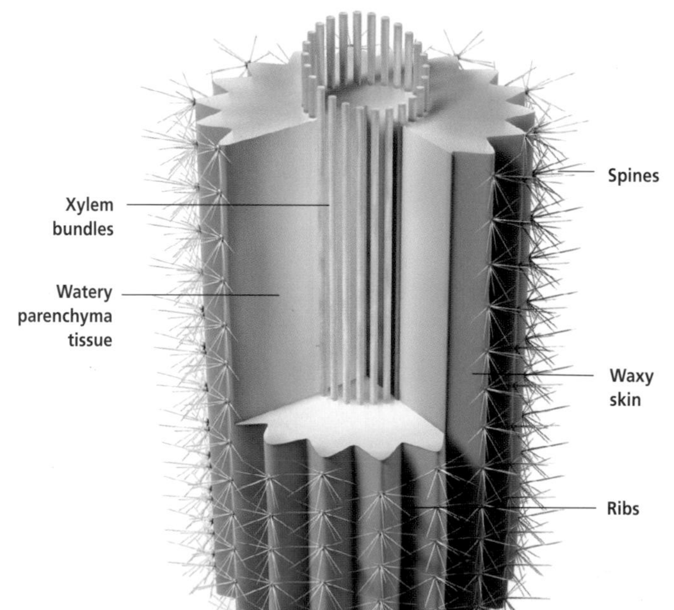

↑ **The Saguaro** (*Carnegiea gigantea*) of the Sonoran Desert of southwestern USA and adjacent Mexico is a stem-succulent, well adapted to survival in this arid landscape.

← **This cross-section of the Saguaro Cactus** reveals the unusual inner structure which is made up of mostly watery tissue with an inner cylinder of xylem strands.

↗ **Cholla** is the name for opuntioid cacti of the genus *Cylindropuntia*. These Jumping Cholla (*C. bigelovii*) shrubs are growing beneath Joshua Trees (*Yucca brevifolia*) in the Joshua Tree National Park, California.

→ ***Euphorbia ingens*** is the Candelabra Tree of eastern Africa. A stem succulent growing as tall as 40 feet (12 m), it is one of 20 or more tree euphorbias which may dominate in some semi-arid regions, such as here in Kenya's Masai Mara Reserve.

Aquatic trees

Trees by their nature cannot be fully aquatic, that is, submerged or floating plants, since such plants do not need trunks to support them and hold them upright. But there are some kinds of trees that normally have their bases in water, either permanently or for much of the time. These are not the same as trees that can tolerate occasional flooding. Aquatic trees can favor either a freshwater or a saltwater environment: those from saltwater are almost by definition mangroves. Some mangroves, though, grow in only slightly saline (brackish) water, such as in upper reaches of estuaries where freshwater streams flow in; and some freshwater aquatic trees will tolerate slight salinity. Trees that grow in fresh or slightly brackish water, such as in rivers, lakes and swamps, are more varied around the world and mostly belong to families or even genera that are characteristic of the flora of their native landmass. They include conifers as well as flowering plants. Among aquatic conifers, the North American Bald Cypress (*Taxodium distichum*) is the best known example, often growing permanently in water 3 feet (1 m) or more deep. A much rarer example is the New Caledonian podocarp *Nageia minor*, a small tree with strikingly conical trunk that grows in quite deep water at edges of rivers.

→ **River Red Gums (*Eucalyptus camuldulensis*)** grow along river banks, on seasonally waterlogged plains and the channels of streams.

↓ **The Bald Cypress (*Taxodium distichum*),** native to southeastern USA, is found in freshwater swamps and on low-lying riverbanks, and tolerates deep water.

↑ **Some types of mangroves,** such as White or Gray Mangroves (above) excrete salt directly through the leaves, while Red Mangroves have roots that totally exclude salt from entering.

↑ **At low tide, mangrove pneumatophores** become exposed. These are rootlike structures that stick up through the soil, much like drinking straws, and enable the intake of oxygen.

ADAPTABILITY OF AQUATIC TREES

It is a significant feature of nearly all the trees that grow with bases covered by fresh or brackish water, that they will grow equally well or sometimes better in properly drained soil, as long as they have access to plentiful soil moisture. Even some mangroves have been successfully grown away from the tidal zone. What this indicates is that such trees tolerate submergence but do not require it. Because they have some intrinsic physiological properties of root or stem tissues that allows them to function in a low-oxygen environment, they have been able to occupy inundated locations where other, potentially competing, trees cannot survive.

MANGROVE ROOT SYSTEMS

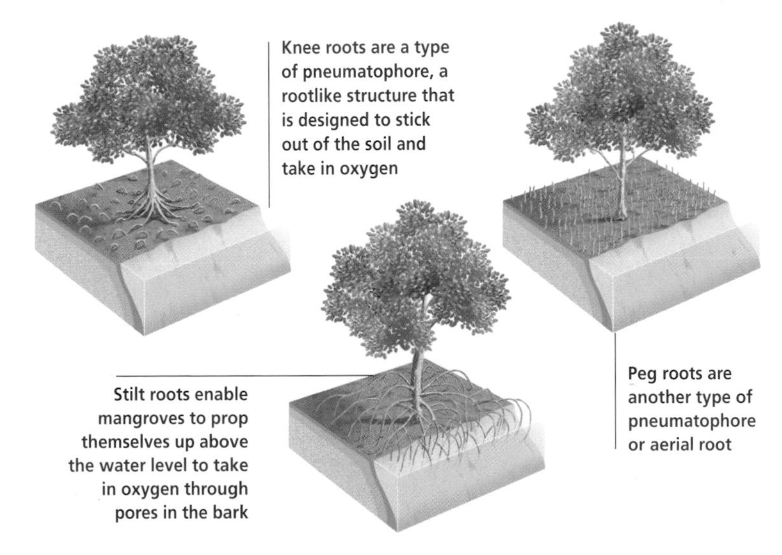

→ **Mangroves and other aquatic trees** have developed various specialized root systems, including knee, peg and stilt roots.

Remarkable trees of the world

The following pages provide a guide to 99 of the world's most exceptional trees. At least a quarter of the world's 300,000-plus flowering plant species, and more than nine-tenths of its 600 conifers, reach tree size. With so many trees to choose from, the species represented here have been included because they have attracted the attention of humans for a variety of reasons, whether economic, cultural, aesthetic or simply botanical. These species range from several of the largest known trees to some that barely count as trees.

Many of the inclusions have become notable tourist attractions, either as individual specimens or in stands—think of the General Sherman *Sequoiadendron* in California, the great *Taxodium mucronatum* at Tule, Mexico, the Joshua Trees (*Yucca brevifolia*) near Los Angeles and some of the huge Kauris (*Agathis australis*) in New Zealand. And of course the Sugar Maples (*Acer saccharum*) of New England in the USA attract countless visitors annually to see their brilliant fall display. However, there are other trees that tourists do not have such ready access to, but are famous for their size, age, rarity or some other feature. The giant Australian Mountain Ash of Tasmania (*Eucalyptus regnans*) and the recently discovered Wollemi Pine (*Wollemia nobilis*) fall into this category.

Each entry in the guide provides key information about a particular tree. This includes the maximum height to which the tree is expected to grow in natural forest conditions; the type of tree and whether it is evergreen or deciduous; its distribution around the world; the habitats where it is usually found; and information about scientific classification, such as the division and family to which the tree belongs. There is always a fascination in discovering the origins of both the botanical and common names of living organisms and these are discussed for many of the trees in this section. A photograph highlights a significant feature of the tree.

A silhouette diagram of each tree indicates overall form and size of a typical specimen.

A human figure (6ft/1.8m) is shown next to the tree to indicate scale.

California Red Fir

Abies magnifica

Height: up to 230 ft (70 m)
Type: evergreen conifer
Occurrence: Sierra Nevada and Klamath Mountains, California, USA; Cascade Ranges, Oregon, USA
Habitat: temperate coniferous forest on steep mountain slopes at 4600–9000 ft (1400–2750 m)
Division: Pinophyta **Family:** Pinaceae

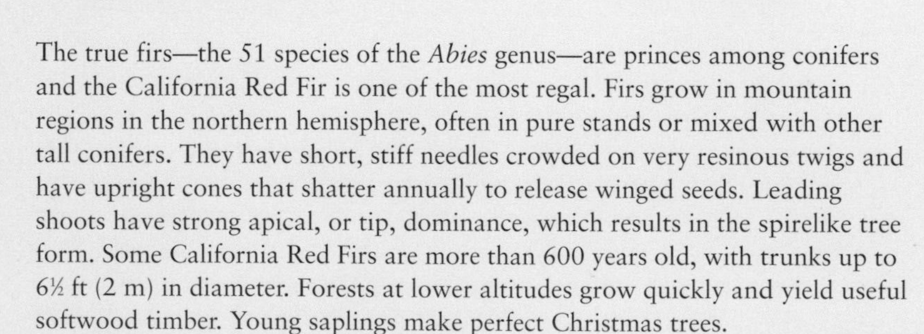

The true firs—the 51 species of the *Abies* genus—are princes among conifers and the California Red Fir is one of the most regal. Firs grow in mountain regions in the northern hemisphere, often in pure stands or mixed with other tall conifers. They have short, stiff needles crowded on very resinous twigs and have upright cones that shatter annually to release winged seeds. Leading shoots have strong apical, or tip, dominance, which results in the spirelike tree form. Some California Red Firs are more than 600 years old, with trunks up to 6½ ft (2 m) in diameter. Forests at lower altitudes grow quickly and yield useful softwood timber. Young saplings make perfect Christmas trees.

The California Red Fir has curved needles, arranged toothbrush-like along the twigs, which are silvery blue when young. The cones, up to 9 in (23 cm) long, are the largest of any fir.

Black Wattle

Acacia mearnsii

Height: up to 60 ft (18 m)
Type: evergreen broadleaf
Occurrence: southeastern Australia (including Tasmania), coast and adjacent ranges up to 2000 ft (600 m) altitude
Habitat: temperate broadleaf forest, mostly colonizing disturbed areas in evergreen eucalypt forest
Division: Magnoliophyta: dicotyledons
Family: Fabaceae (Leguminosae)

What makes the Black Wattle stand out among the 950 Australian *Acacia* species is its use in commercial forestry in southern and eastern Africa, India, Brazil and Chile. Fast-growing and adapted to poor soils, it is a major source of tanbark for the leather and adhesives industries. The dark brown bark, stripped from felled trees, is chipped and exported. It is rich in tannins that preserve and strengthen animal hides and also has uses in plastics and plywood gluing. The spent tanbark is laid on horse-racing and athletics tracks. Black Wattle timber is used for firewood, poles, tool handles and other small items.

The Black Wattle is one of a group of Australian wattles with feathery leaves made up of many tiny leaflets. A profusion of ball-shaped, cream flowers appear in late spring.

Babul

Acacia nilotica

Height: up to 40 ft (12 m)
Type: dry-season-deciduous broadleaf
Occurrence: throughout most of Africa; Arabia; southern Asia east to Burma
Habitat: savanna and semiarid scrubland, mostly on plains, soils from sand to heavy clay
Division: Magnoliophyta: dicotyledons
Family: Fabaceae (Leguminosae)

Acacia, as currently understood, comprises about 1350 species from all the warmer parts of the world. The most wide-ranging of these is the Babul. A thorny tree characteristic of semiarid scrub and savanna, its foliage and pods are browsed by herbivores such as giraffes. Its human uses are multiple. The gum from the bark is one source of commercial gum arabic, used in pharmaceutical products, printing inks and adhesives. The tough wood makes tool handles, fence posts and firewood. Bark and pods are used for tanning leather, and the tannin has been used to make ink. A range of medicinal uses is also reported.

The branches have paired spines, and the small, bipinnate leaves have crowded, tiny leaflets. Balls of golden yellow, scented flowers appear in summer, followed by long, beaded pods.

Umbrella Thorn

Acacia tortilis

Height: up to 50 ft (15 m)
Type: dry-season-deciduous broadleaf
Occurrence: Africa south of the Sahara; Arabia; southwest Asia
Habitat: savanna and scrubland, on plains and dry watercourses, mostly on clay soils
Division: Magnoliophyta: dicotyledons
Family: Fabaceae (Leguminosae)

A striking feature of the veld country of southern Africa are the flat-topped "thorn trees"—as many as 40 species of *Acacia*. Some, such as the Umbrella Thorn, have much wider distributions. They are not closely related to most Australian acacias, and many botanists now believe they should be a separate genus. The Umbrella Thorn has small, bipinnate leaves. The twigs bear pairs of long, straight spines or short, hooked spines at each node. Clustered, globular heads of cream flowers in summer are followed by coiled pods. Fast-growing and drought-resistant, it is used in India for firewood and to combat soil erosion.

The tree's broad crown gave rise to its common name, the Umbrella Thorn. Its foliage is highly nutritious and preferred by giraffes, who browse the leaves despite the fierce spines.

Japanese Maple

Acer palmatum

Height: up to 40 ft (12 m)
Type: winter-deciduous broadleaf
Occurrence: Japan; Korea; eastern China
Habitat: temperate deciduous broadleaf forest on hills and mountain slopes
Division: Magnoliophyta: dicotyledons **Family:** Aceraceae

The chief claim to fame of this small tree is its wide cultivation for the beauty of its foliage, and the extraordinary number of cultivars (garden varieties) that have been raised from the one species—more than 400 have been named and new ones appear in catalogs every year. Cultivars vary in size, growth form, leaf shape and coloring, with both spring and fall color being a feature. For centuries before this maple was introduced to the West in 1820, numerous cultivars had been selected and named in Japan. The species name, *palmatum,* refers to the palmate (handlike) lobing of the leaves, with five to seven narrow lobes. These in turn are toothed or, in many cultivars, further divided into fine lobes. Cultivars are perpetuated by grafting onto seedling rootstocks.

With its gnarled trunk and fine fall color, an old specimen of Japanese Maple makes a spectacular display in Tokyo's famous Shinjuku Gyoen Park.

Sugar Maple

Acer saccharum

Height: up to 120 ft (37 m)
Type: winter-deciduous broadleaf
Occurrence: eastern and central USA; southeast Canada
Habitat: temperate deciduous forest, mostly on hills or well-drained soils in valleys
Division: Magnoliophyta: dicotyledons **Family:** Aceraceae

This is arguably the best known of all North American trees, the one whose leaf symbolizes Canada and whose fall color sets the hills of New England ablaze. It is the source of that most famous of American sweeteners, maple syrup, and yields a valuable timber too. The maple genus (*Acer*) has about 120 species, distributed mainly through temperate regions of the northern hemisphere, most in eastern Asia. In the Americas there are only 12 species, most of them smaller trees than the Sugar Maple. All maples have leaves arranged in opposite pairs. In most, the leaves are lobed or deeply toothed and about the same width and length. Flowers are mostly small and greenish, and the two-winged fruits, or "keys," are a constant feature.

In New England, fall color peaks in October. Leaves turn a different shade on individual trees, from pale yellow to deep red. Maple syrup is collected from late winter to early spring.

Baobab

Adansonia digitata

Height: up to 72 ft (22 m)
Type: dry-season-deciduous broadleaf
Occurrence: African tropics; from the Sahara to north of Johannesburg, South Africa
Habitat: savanna and arid scrubland, in deep, sandy soil or on rocky outcrops
Division: Magnoliophyta: dicotyledons
Family: Malvaceae

If any tree symbolizes Africa, it is this. Renowned for its vast bulk—trunk up to 36 ft (11 m) in diameter, crown spread to 120 ft (37 m)—one Baobab can contain 120 tons of water in its spongy wood. Elephants sometimes rip trees apart in the dry season and chew the wood for its water. Large herbivores also relish the foliage and fallen fruit, full of seeds in a sweet–sour pulp. The seeds are then deposited in heaps of dung, which aids germination and seedling growth. The many human uses of the Baobab include ropes and cord from the bark fiber. The *Adansonia* genus is noted for its broken distribution: one species in Africa, six in Madagascar and one in northwestern Australia.

Baobabs are leafless until the wet season starts. Leaves are palmately divided into five to nine leaflets. Large, fragrant white flowers hang on long stalks and are pollinated by bats.

Horse Chestnut

Aesculus hippocastanum

Height: up to 100 ft (30 m)
Type: winter-deciduous broadleaf
Occurrence: mountains of central Balkans and Caucasus
Habitat: temperate deciduous forest
Division: Magnoliophyta: dicotyledons
Family: Hippocastanaceae

When this tree reached western Europe in the early 17th century, it became popular immediately as an outstanding ornamental for parks and large estates. But its origin remained unknown for the next 250 years, until it was found growing wild in remote valleys in northern Greece and Albania. Even now, its exact wild distribution is uncertain, but it is likely that it extends east to northern Iran. Europe is a western outpost of the *Aesculus* genus, which has five species in the China–Himalaya region and seven in North America. The American species are known as "buckeyes" for their large brown seeds, held in a leathery capsule. The similar seeds of the Horse Chestnut are the "conkers" (conquerors) that British children use in a traditional game.

In late spring the broad canopy is decked with foot-tall panicles of white flowers with long stamens, their frilled petals blotched pink and yellow. Each leaf has five to seven leaflets.

Kauri

Agathis australis

Height: up to 180 ft (55 m)
Type: evergreen conifer
Occurrence: northern half of North Island, New Zealand
Habitat: subtropical rain forest on deep, fertile soil, often in boggy ground
Division: Pinophyta **Family:** Araucariaceae

One of the world's most impressive conifers, the New Zealand Kauri once grew in vast forests that covered as much as 3 million acres (1.2 million hectares) of the North Island lowlands. Through much of the 19th century these forests were ruthlessly exploited, the tough, durable timber at first prized for the spars of sailing ships, then later for furniture and general construction. Trees 10 ft (3m) or more in diameter, with straight trunks free of branches for over 100 ft (30 m), were felled and sawed. Then in the 1880s large areas of the remaining forests were destroyed by wildfire. Now only a few thousand acres of high-quality Kauri forest remain, in highly protected reserves. The largest trees are major tourist attractions.

The *Agathis* genus dates back to the Jurassic period, around 145 million years ago. Its 20 species are distinct among conifers for broad, leathery leaves and large, spherical cones.

Quiver Tree

Aloe dichotoma

Height: 15–25 ft (5–8 m)
Type: succulent evergreen swordleaf
Occurrence: Namibia; Northern Cape, South Africa
Habitat: savanna and scrubland, on gravelly plains and stony hills
Division: Magnoliophyta: monocotyledons
Family: Asphodelaceae

Aloe is a large genus of succulent plants. Most of the 400 or so species are native to Africa, spilling over into Arabia, but many also come from Madagascar. They vary from tiny, ground-hugging or grasslike plants through shrubs and scrambling climbers to medium-size trees. The tree aloes have a characteristically candelabrum-like growth habit, with each branch ending in a rosette of fleshy, prickly edged swordleaves. The Quiver Tree is distinctive for its fat trunk covered in yellowish plates of bark with sharp, upturned edges. The thick branches were hollowed out and used as quivers by the San people (Bushmen), hence the common name. In winter the tree bears spikes of nectar-rich yellow flowers, attracting birds, insects and even baboons.

The view up through the branches of a Quiver Tree shows the early development of its characteristic sharp-edged bark plates. They become more prominent on the thick trunk.

Monkey Puzzle

Araucaria araucana

Height: up to 130 ft (40 m)
Type: evergreen conifer
Occurrence: central Chile; adjacent Andean slopes of Argentina
Habitat: temperate coniferous forest or broadleaf evergreen forest, mostly on steep mountain slopes
Division: Pinophyta **Family:** Araucariaceae

A tree of striking form, the Monkey Puzzle is a member of an ancient plant family from the Cretaceous period. Its perfectly symmetrical crown, umbrella-like with age, is a tangle of thick, curved branches clothed in overlapping, sharp-pointed leaves that are amazingly stiff and hard. Also remarkable is the longevity of individual leaves, which can persist, green and alive, attached to the trunk base of a 25-year-old tree. The tree was known to early Spanish settlers in Chile, but was only introduced to Europe after botanist–explorer Archibald Menzies pocketed some of its nuts from the Chilean Governor's dining table in 1795; from these, five trees were raised in England.

Proving well adapted to the English climate, the Monkey Puzzle Tree (the name given by an astonished English viewer) was all the rage as a garden ornament by the mid-19th century.

Bunya Bunya

Araucaria bidwillii

Height: up to 160 ft (49 m)
Type: evergreen conifer
Occurrence: northeast and southeast Queensland, Australia
Habitat: tropical and subtropical rain forest on fertile soils
Division: Pinophyta
Family: Araucariaceae

Strong evidence that this majestic conifer is a relic of much older vegetation is provided by its isolated occurrence in two limited areas of rain forest 800 miles (1300 km) apart, almost at opposite ends of the very large state of Queensland. And in fact there are many fossils that show that these trees are the only survivors of a distinct type of *Araucaria* that was widespread in various parts of the world as far back as the Jurassic period. The *Araucaria* genus has 18 living species, the majority confined to New Caledonia, but with two each in Australia and South America. Aborigines of southeast Queensland knew this tree as *Bunya Bunya* and traveled long distances every year to feast on its large, nutritious seeds, contained in very large cones.

The Bunya Bunya grows most prolifically in the Bunya Mountains, named for the tree. Widely spaced climbing notches can still be seen in the bark of some older trees there.

Pacific Madrone

Arbutus menziesii

Height: up to 100 ft (30 m)
Type: evergreen broadleaf
Occurrence: west coast of North America, from southwest British Columbia, Canada, to northern Baja California, Mexico
Habitat: coniferous forest to evergreen woodland
Division: Magnoliophyta: dicotyledons **Family:** Ericaceae

Early Spanish colonists of North America recognized the affinity of this lovely tree to the Mediterranean *madroño* (*A. unedo*). Their English-speaking successors modified the name to Madrone or Madrona. Most abundant in northern California, from the coast to the Sierra foothills, the Pacific Madrone grows mostly on rocky ground in a range of forest and woodland. Its most striking feature is its bark—smooth and rich orange brown for much of the year, but peeling dramatically in summer to reveal pale greenish cream new bark, cool and clammy to the touch. The *Arbutus* genus has eight species in North America and the Mediterranean. All bear clusters of white to pinkish, bell-shaped flowers in spring and wrinkled, edible but acid fruits in fall.

The Pacific Madrone is a great survivor, withstanding exposed sites, poor soils and forest fires, as seen in this scarred specimen. It is growing with Ponderosa Pine (*Pinus ponderosa*).

Neem

Azadirachta indica

Height: up to 60 ft (18 m)
Type: dry-season-deciduous broadleaf
Occurrence: tropical and subtropical Asia; origin uncertain, possibly Burma
Habitat: tropical deciduous forest, mostly in disturbed areas in soils of medium to high fertility
Division: Magnoliophyta: dicotyledons **Family:** Meliaceae

This is one of those trees that has been promoted as a "miracle tree," and in truth it has a remarkable range of uses, most of them first discovered in India, where it has long been cultivated and valued. The scientific name *Azadirachta* was taken from the Persian *azad-dirakht,* but its application to Neem may have resulted from confusion with the similar Persian Lilac, *Melia azedarach*. Neem (or *nim*) is its Urdu and Hindi name. Its most important use is as a pesticide. Both the foliage and seed oil are toxic to a wide range of insects, as well as pests such as ticks and nematodes, but relatively harmless to mammals. Pests do not seem to develop resistance to the active compound, azadirachtin.

Neem has pinnate leaves with curved leaflets, while the Persian Lilac's are bipinnate. White flowers occur with the new foliage flush, followed by yellow drupes with a large oily seed.

Paper Birch

Betula papyrifera

Height: up to 80 ft (24 m)
Type: winter-deciduous broadleaf
Occurrence: northern USA (including Alaska) and Canada, north to the Arctic Circle
Habitat: temperate deciduous broadleaf and boreal forest
Division: Magnoliophyta: dicotyledons **Family:** Betulaceae

The birches (*Betula* genus), with about 35 species, include the northernmost of all broadleaf trees. They were foremost among the trees that reinvaded the bare, rocky soils exposed as the vast continental ice sheets melted and retreated at the end of the last Ice Age, about 10,000 years ago. The rapid spread of birches is aided by their huge production of very light, wind-blown seed. Also, their abundant, wind-carried pollen needs no insects for effective fertilizing of the flowers. The Paper Birch, one of the more northern American species, has beautiful white, papery bark, famous for its use in birch-bark canoes as well as for roofing of dwellings and even drinking cups. The sappy inner bark provides winter food for moose and deer.

Paper Birch bark develops in layers. The outermost layers peel off to reveal subtle variations in color. It contains the water-repellent compound suberin, which makes it waterproof.

Frankincense

Boswellia sacra

Height: up to 20 ft (6 m)
Type: dry-season-deciduous broadleaf
Occurrence: Yemen and Oman, southern Arabian Peninsula; Somalia, Horn of Africa
Habitat: semiarid scrubland on stony hills
Division: Magnoliophyta: dicotyledons **Family:** Burseraceae

The product of this tree, a dried gum resin, is far better known than the tree itself. Frankincense derives from Old French for "pure incense." Its older Latin name is *olibanum,* from the Arabic *al-luban,* for "milk," a reference to the tree's milky sap, which, when tapped and dried, becomes the incense. Like some other high-value products used since ancient times, frankincense passed through many hands along trade routes and its source was something of a mystery. It was not until 1867 that Swiss pharmacist Friedrich Flückiger published the scientific name, *Boswellia sacra,* for this small, strongly resinous tree from the stony hills of southern Arabia. He placed it in the genus named earlier in honor of the famous James Boswell, friend of Dr. Johnson.

Leafless in the dry season, Frankincense is one of about 20 *Boswellia* species from the drier parts of Africa and southern Asia. They have pinnate leaves with bluntly toothed leaflets.

Paper Mulberry

Broussonetia papyrifera

Height: up to 50 ft (15 m)
Type: winter-deciduous broadleaf
Occurrence: China; hills of Southeast Asia; long cultivated in Pacific Islands
Habitat: deciduous broadleaf forest and rain forest, in valleys and ravines
Division: Magnoliophyta: dicotyledons **Family:** Moraceae

The Paper Mulberry has been planted for so long in all parts of China and Southeast Asia that its native region is uncertain. It is a spreading tree, easily grown—in some places it is a troublesome weed. In some Pacific Islands it is the famed source of tapa cloth, used for traditional handicrafts and artworks, and made by beating and felting the inner bark fiber with a wooden club. But its use as a fiber for cordage, weaving and papermaking is much older in eastern Asia, which is probably why early Polynesian seafarers first carried the plants to their new island homes. This tree has an unusually wide climatic tolerance, thriving anywhere from the cold steppes of northern China to tropical Fiji.

The Paper Mulberry has large, rough-surfaced leaves, often lobed on young plants. Flowers are different sexes on different plants; the red flowers in these globular heads are female.

Chinese Camellia

Camellia reticulata

Height: up to 50 ft (15 m)
Type: evergreen broadleaf
Occurrence: mountains of Yunnan, southwest Sichuan and western Guizhou, in southwest China
Habitat: deciduous broadleaf forest
Division: Magnoliophyta: dicotyledons **Family:** Theaceae

Not only is this one of the largest trees of the 120 or so species of *Camellia* (almost 100 of them Chinese), it also has the largest and showiest flowers—at least in its selected garden forms. Some of these have been grown in the grounds of Buddhist monasteries in Yunnan since the 10th century or even earlier. Some massive specimens are known to be over 500 years old, such as the Ten Thousand Flower Camellia in a monastery near Lijiang. They were unknown in the West until a Captain Rawes of the East India Company brought one variety from Canton to London in 1820. A sensation when it first flowered in 1826, it is still widely grown under the cultivar (garden variety) name of 'Captain Rawes.'

Cultivars of the Chinese Camellia have multipetaled, or "double," flowers up to 9 in (23 cm) across, in contrast to the smaller, wild forms with five- to seven-petaled flowers.

Ylang Ylang

Cananga odorata

Height: up to 50 ft (15 m)
Type: evergreen broadleaf
Occurrence: tropical Southeast Asia to far northern Australia
Habitat: tropical lowland rain forest and vine thickets
Division: Magnoliophyta: dicotyledons
Family: Annonaceae

A member of the large Custard Apple family of tropical trees and climbers, Ylang Ylang is renowned for the deliciously sweet scent of its flowers. In fact, the essential oil of Ylang Ylang is an important ingredient in some of the world's great perfumes, including the famous No. 5 formulated by Chanel's Ernest Beaux around 1920. The oil is precious, a single gram of it being produced from distilling about 1 lb (500 g) of fresh flowers, laboriously picked from the tree. The exotic-sounding common name, Ylang Ylang, is the tree's name in Tagalog, the main language of the Philippines, which was the major original source of the oil. Belonging to a genus of only two species, the Ylang Ylang is also a popular ornamental tree in the tropics.

The flowers' features indicate Ylang Ylang's early origin among flowering plants. The scent attracts certain beetles, which appeared early in insect evolution, to ensure pollination.

Saguaro

Carnegiea gigantea

Height: up to 50 ft (15 m)
Type: leafless succulent
Occurrence: southern California and Arizona, USA; Sonora, Mexico
Habitat: arid scrubland on plains and low, stony hills
Division: Magnoliophyta: dicotyledons **Family:** Cactaceae

If any plant is a symbol of the Arizona deserts, it is this, the archetypal cactus with upturned "arms." It is one of the more massive members of the cactus family, weighing up to 2 tons with a trunk up to 3¼ ft (1 m) in diameter. But it is only one of many tree cacti, most of which come from Mexico and South America, and some of these can grow even larger, with many more branches. Like most other cacti, the Saguaro is leafless, with photosynthesis taking place in the green stem. It is the only member of the *Carnegiea* genus, named in 1908 in honor of the philanthropist and patron of science Andrew Carnegie. The common name, Saguaro (suh-WAH-ro), comes from the plant's Native American name via early Spanish settlers.

Rising above the desert, Saguaro have strikingly ribbed stems. The ribs close up concertina-like to reduce the stem's volume as the plant's water reserves are used up during drought.

Shagbark Hickory

Carya ovata

Height: up to 100 ft (30 m)
Type: winter-deciduous broadleaf
Occurrence: USA east from Kansas; far southeast Canada
Habitat: broadleaf forest on alluvial soils in bottomlands and along stream banks
Division: Magnoliophyta: dicotyledons **Family:** Juglandaceae

Closest cousins to the walnuts (*Juglans* genus), the hickories, pignuts, pecan and mockernut make up the *Carya* genus. The 17 species are mainly North American but with a few from eastern Asia. Like the walnuts, they have oil-rich, nutritious nuts—bitter in some species—and pinnate leaves. The leaves and leaflets of vigorous young hickory saplings are usually quite large. Hickory timber is known for two reasons: its strength and resilience make it ideal for tool handles, and the distinctive tang of its smoke gives the best flavor to smoked meats. Shagbark Hickory has very striking bark, which peels off in long, thick plates, each plate curving outward at both top and bottom and giving the trunk a coarse, shaggy look.

With its fall foliage, the Shagbark Hickory would be more popular in parks and gardens if its growth was not so slow and it was easier to transplant. The edible nuts are sweet.

Beach Casuarina

Casuarina equisetifolia

Height: up to 80 ft (24 m)
Type: evergreen broadleaf
Occurrence: tropical Australia; southern Asia; Pacific and Indian Ocean Islands
Habitat: on beaches and in disturbed forest behind beaches, mainly on sand
Division: Magnoliophyta: dicotyledons **Family:** Casuarinaceae

This tree is only a "broadleaf" in the botanical sense. At first glance all of the casuarinas look like conifers, with apparent needles, no obvious flowers and seeds borne in cones. But the "needles" are twigs with whorls of tiny leaves fused to their faces. The flowers are present, although small, unisexual and wind-pollinated. The "cones" are actually dense clusters of small, woody bracts that split at maturity to release tiny, winged fruits. New evidence shows they are most closely related to the birches and alders of the northern hemisphere, similarly wind-pollinated trees with wind-borne fruits. Like alders, their roots associate with the actinobacterium *Frankia,* which can fix nitrogen from air.

Beach Casuarina is widely planted in the tropics and subtropics for firewood, shelter belts and erosion protection. It is a weed tree known as Australian Pine in southern Florida.

Cedar of Lebanon

Cedrus libani

Height: up to 140 ft (43 m)
Type: evergreen conifer
Occurrence: Lebanon; Syria; southern Turkey
Habitat: temperate coniferous forest on mountain slopes at 4000–7000 ft (1220–2130 m)
Division: Pinophyta **Family:** Pinaceae

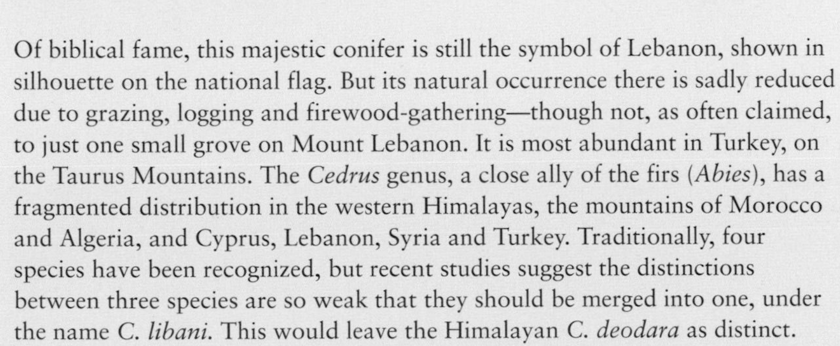

Of biblical fame, this majestic conifer is still the symbol of Lebanon, shown in silhouette on the national flag. But its natural occurrence there is sadly reduced due to grazing, logging and firewood-gathering—though not, as often claimed, to just one small grove on Mount Lebanon. It is most abundant in Turkey, on the Taurus Mountains. The *Cedrus* genus, a close ally of the firs (*Abies*), has a fragmented distribution in the western Himalayas, the mountains of Morocco and Algeria, and Cyprus, Lebanon, Syria and Turkey. Traditionally, four species have been recognized, but recent studies suggest the distinctions between three species are so weak that they should be merged into one, under the name *C. libani.* This would leave the Himalayan *C. deodara* as distinct.

This large specimen exhibits the characteristic low-branched trunk and ascending limbs. Introduced to England in about 1650, there are several centuries-old trees still growing.

Kapok Tree

Ceiba pentandra

Height: up to 230 ft (70 m)
Type: dry-season-deciduous broadleaf
Occurrence: tropical Africa and America; introduced to tropical Asia
Habitat: tropical rain forest and vine thickets, usually in valley bottoms in fertile soils
Division: Magnoliophyta: dicotyledons **Family:** Malvaceae

This is one of the notable trees of the tropics because it is the major source of kapok, a cotton wool-like plant fiber now largely replaced by synthetics. Also, it is possibly Africa's tallest native tree. The Kapok Tree is thought to have been taken from Africa to Madagascar, then Indonesia and elsewhere in Asia, as much as 1500 years ago, no doubt for its useful fiber and edible young fruit and shoots. More recently, kapok was used for mattresses, pillows, life jackets and refrigerator insulation. Just like the cotton plant, which belongs to the same broadly defined family, the fibers are the hairs on the surface of the oil-rich seeds, tightly packed within a pod that bursts open at maturity.

The Kapok Tree is fast-growing with a straight, smooth-barked trunk. The cream flowers appear in the dry season; bat-pollinated, they are followed by long, pendent pods.

Judas Tree

Cercis siliquastrum

Height: up to 40 ft (12 m)
Type: winter-deciduous broadleaf
Occurrence: eastern Mediterranean; Middle East
Habitat: temperate deciduous forest and semiarid scrubland
Division: Magnoliophyta: dicotyledons
Family: Fabaceae (Leguminosae)

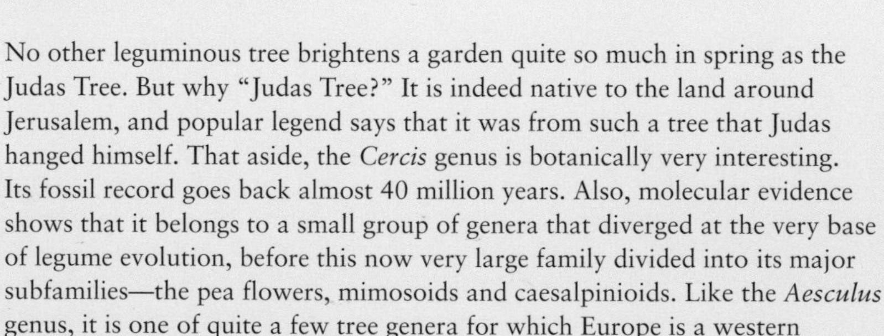

No other leguminous tree brightens a garden quite so much in spring as the Judas Tree. But why "Judas Tree?" It is indeed native to the land around Jerusalem, and popular legend says that it was from such a tree that Judas hanged himself. That aside, the *Cercis* genus is botanically very interesting. Its fossil record goes back almost 40 million years. Also, molecular evidence shows that it belongs to a small group of genera that diverged at the very base of legume evolution, before this now very large family divided into its major subfamilies—the pea flowers, mimosoids and caesalpinioids. Like the *Aesculus* genus, it is one of quite a few tree genera for which Europe is a western outpost, most of its species being North American or east Asian.

The Judas Tree flowers on bare branches in midspring. The flowers resemble a typical pea flower but show significant differences in structure. Almost circular leaves follow.

Floss Silk Tree

Chorisia speciosa

Height: up to 100 ft (30 m)
Type: dry-season-deciduous broadleaf
Occurrence: southwest Brazil; northern Argentina; Paraguay; Bolivia; southern Peru
Habitat: savanna and drier hill forest, mostly under 3000 ft (910 m) altitude
Division: Magnoliophyta: dicotyledons **Family:** Malvaceae

One of the most beautiful tropical trees, the Floss Silk Tree makes a profuse display of large pink flowers during the early tropical dry season or from early fall to early winter. The flowers, 4–6 in (10–15 cm) wide, begin opening as the last leaves fall and continue on leafless branches. Each of the five large petals has a cream base streaked with purple brown, changing to bright pink on the outer rim. Petal color and width vary subtly from tree to tree. The fruit that follows is football-shaped, up to 8 in (20 cm) long and contains seeds tightly packed in kapok, just as in the *Ceiba* genus. In fact, recent botanical opinion has it that *Chorisia* cannot be distinguished as a genus from *Ceiba*.

A young Floss Silk Tree's trunk is covered in conical prickles, often lost as the tree ages. It is a popular ornamental in warm-temperate places such as California and eastern Australia.

Cinnamon

Cinnamomum verum

Height: up to 40 ft (12 m)
Type: evergreen broadleaf
Occurrence: southern India; Sri Lanka
Habitat: tropical rain forest; cultivated in areas of high rainfall, in fertile soil
Division: Magnoliophyta: dicotyledons **Family:** Lauraceae

The botanical name means "true cinnamon," because the bark of a number of other species of the large *Cinnamomum* genus is either substituted for, or mixed with, the bark of this tree, the most highly valued species. This famous spice was known to the ancient Greeks, who obtained it from Phoenician traders, though its botanical and geographic origins were probably unknown to both. The name comes from the Greek *kinnamon,* in turn believed to have come from a Semitic language. In its native lands its use has continued for millennia; for example, as a major ingredient of the spice mix garam masala. The tree belongs to the large, mainly tropical laurel family, most members of which have aromatic oils in their bark and leaves.

In plantations, the trees are cut back to encourage vigorous stems. These are harvested, the outer bark removed and inner bark peeled and dried for the familiar cinnamon "quills."

Coconut Palm

Cocos nucifera

Height: up to 100 ft (30 m)
Type: feather palm
Occurrence: tropical Indian and Pacific Ocean Islands; exact origin uncertain
Habitat: sandy seashores, but cultivated in lowlands throughout tropics
Division: Magnoliophyta: monocotyledons
Family: Arecaceae (Palmae)

It may be an advertising cliché for beach resorts, but the Coconut Palm is one of the tropical world's most important economic plants. It is highly productive on sites where few if any crops will grow, such as coral sands on hurricane-lashed atolls. Almost every part has a use, from the edible terminal bud ("millionaire's salad") to the trunks of old trees (for house timbers and water pipes). And the large fruit, borne year-round, has a hard-shelled nut inside a thick, fibrous husk. The nut's "meat" and "milk" are mainstays of tropical Asian cuisine. The extracted oil and husk fiber (coir) are world commodities.

Coconut Palms on beaches such as this often grow from coconuts carried by ocean currents and cast up by storm waves. The seed remains viable in seawater for up to 2 weeks.

Talipot Palm

Corypha umbraculifera

Height: up to 60 ft (18 m)
Type: fan palm
Occurrence: tropical mainland Asia; exact origin uncertain due to spread by humans
Habitat: disturbed forest and scrub on alluvium in lowland valleys and plains
Division: Magnoliophyta: monocotyledons
Family: Arecaceae (Palmae)

The most dramatic features of the Talipot are the massive size of its fronds and its mode of flowering: only after 20 to 40 years of vigorous growth does it produce a gigantic terminal panicle that bears as many as one million flowers. As the subsequent huge crop of 1½-in (4-cm) fruits ripens, the whole palm slowly dies and eventually topples over. It is not the only palm with this mode of growth, but it is the largest. Early European observers in India recounted tales of how a single frond—which can be up to 10 ft (3 m) wide—could be used as an umbrella, providing shelter for 15 to 20 people.

In the late flowering stage, the Talipot Palm's fronds die below the giant panicle. The palm has many uses, including trunk starch for sago and leaf segments for writing materials.

Cannonball Tree

Couroupita guianensis

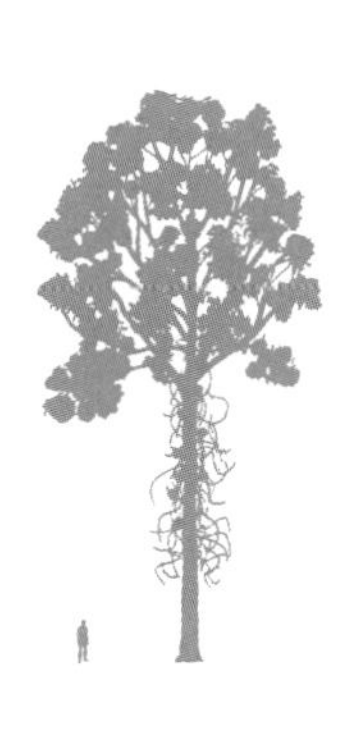

Height: up to 120 ft (37 m)
Type: evergreen broadleaf
Occurrence: South America, from Panama to the Amazon
Habitat: lowland rain forest on deep, alluvial soils
Division: Magnoliophyta: dicotyledons
Family: Lecythidaceae

You only need to look at the tree to know why it got the name of Cannonball Tree. The fruits, 8 in (20 cm) in diameter, are hard and woody, and contain an unpleasant-smelling pulp in which the seeds are embedded. The fruits appear after the curious and colorful large flowers, which are produced continuously on flowering branches that emerge directly from the lower trunk and grow longer every year. This habit of flowering from the trunk is known as "cauliflory," and the Cannonball Tree is a prime example. It belongs to the same plant family as that other famous tree of the Amazon, the Brazil Nut. Large, fleshy flowers with a complex structure and innumerable stamens, pollinated by bees or bats, are a family feature.

The Cannonball Tree intrigues visitors to tropical parks and gardens. In the wild, the fruits are eaten by peccaries, who crack them open to eat the pulp and thus disperse the seeds.

Calabash Tree

Crescentia cujete

Height: up to 30 ft (10 m)
Type: dry-season-deciduous broadleaf
Occurrence: West Indies; Central and South America, from central Mexico to Bolivia
Habitat: savanna, vine thickets and disturbed woodland
Division: Magnoliophyta: dicotyledons **Family:** Bignoniaceae

The name "Calabash" is ambiguous. It can refer either to this small tree in the bignonia family or to the Bottle Gourd (*Lagenaria siceraria*), a climber in the pumpkin family. It is chiefly from the latter that the calabashes, or maracas, of Latin and African music are made. The Calabash Tree has hard-shelled fruits, 6–12 in (15–30 cm) in diameter, that contain a sour–sweet edible pulp in which numerous seeds are embedded. The fruits are preceded by trumpet-shaped, richly scented, cream to pinkish brown flowers, apparently growing directly out of the thick lower branches. It is thought that the position of the flowers below the foliage canopy helps them to be found more readily by bats using echolocation, who then feed on the nectar and pollinate the flowers.

The Calabash Tree is cauliflorous. The dried, hollowed-out fruits have many traditional uses, such as food and water containers, carved ornaments and musical instruments.

Mediterranean Cypress

Cupressus sempervirens

Height: up to 110 ft (34 m)
Type: evergreen conifer
Occurrence: southern Greece; southern Turkey; Libya; Cyprus; Syria; northern Iran
Habitat: temperate coniferous forest, woodland and scrubland on rocky hill slopes and in ravines
Division: Pinophyta **Family:** Cupressaceae

Also known as the Italian Cypress and Funeral Cypress because of its long association with Italian landscapes and cemetery plantings, this cypress is now thought to have originated farther east. The very narrow, columnar shape is not genetically fixed but has been selected for over centuries of planting. When grown from seed, some trees show the ancestral form—a broader, more horizontally branched habit. This is the only European representative among 20 *Cupressus* species, half of them native to the western USA and Mexico, the remainder to eastern Asia, the Himalayas and the Mediterranean. Molecular studies now show the American species are only distantly related to the others.

The Mediterranean Cypress has been a favorite in formal landscapes since ancient Roman times. The Romans valued the durable, close-grained wood, but it is seldom available now.

Black Tree Fern

Cyathea medullaris

Height: up to 60 ft (18 m)
Type: tree fern
Occurrence: New Zealand; Fiji; Tahiti; Marquesas and Austral Islands
Habitat: tropical and subtropical rain forest and scrubland, often colonizing disturbed areas
Division: Pteridophyta: Pteridopsida **Family:** Cyatheaceae

One of the world's tallest tree ferns, the Black Tree Fern grows abundantly in most parts of New Zealand. The crown on a vigorous specimen can be massive, with fronds up to 20 ft (6 m) long and 6½ ft (2 m) wide. The trunk is covered in large scars left where old fronds have fallen off. As with all tree ferns, it contains no wood, only a skeleton of very tough, leathery fiber bands distributed through a mass of soft, sappy storage tissue. The fiber provides strength in tension, the storage tissue gives resistance to compression. The lower trunk is clothed in a dense sheath of aerial roots, dead on the outside, which give the plant extra strength and stability.

The underside of fronds bear thousands of spore patches, each with numerous spore sacs. These split to shed dustlike spores—millions from a single fern are carried far by the wind.

Dove Tree

Davidia involucrata

Height: up to 80 ft (24 m)
Type: winter-deciduous broadleaf
Occurrence: mountains of central China
Habitat: temperate deciduous broadleaf forest on steep mountain slopes
Division: Magnoliophyta: dicotyledons **Family:** Nyssaceae

Few Chinese trees caused as much excitement and admiration as the Dove Tree did when it was first introduced to the West in about 1900. The pairs of large white bracts, up to 8 in (20 cm) long, that enfold the pendent globular heads of small flowers, give the tree a unique appearance when blossoming in late spring. This species had been discovered in 1869 by the renowned French missionary-naturalist Father Armand David, and specimens he sent to Paris were classified as a new genus and named in his honor. But it was not until the end of the century that seeds were obtained and sent to France, although only one seedling from the first batch survived. Soon afterward, however, large seed collections were sent to England and France and cultivated successfully.

Flower heads on slender stalks have many dark purplish, male flowers and usually only one green female flower. The contrasting large white bracts may help guide pollinating insects.

Flamboyante

Delonix regia

Height: up to 60 ft (18 m)
Type: dry-season-deciduous broadleaf
Occurrence: western Madagascar; widely cultivated and naturalized in tropics
Habitat: tropical deciduous forest, savanna and scrubland on rocky hillsides
Division: Magnoliophyta: dicotyledons
Family: Fabaceae (Leguminosae)

One of the most spectacular flowering trees of the tropics, this is known by a variety of common names. Poinciana or Royal Poinciana come from its former classification under the genus *Poinciana,* while in India it is called *gul mohur* (golden flower). In its native Madagascar, it is now endangered in the wild due to habitat loss and burning of the wood for charcoal. In some countries it is becoming naturalized and is potentially a troublesome weed. The flowers are individually large and very profuse, with four scarlet petals and a slightly longer petal streaked with white, orange and purple. A yellow form also exists.

Flamboyante has an umbrella-like canopy that bursts into scarlet blossom at the end of the tropical dry season. It continues flowering into the wet season as the ferny foliage appears.

Dragon Tree

Dracaena draco

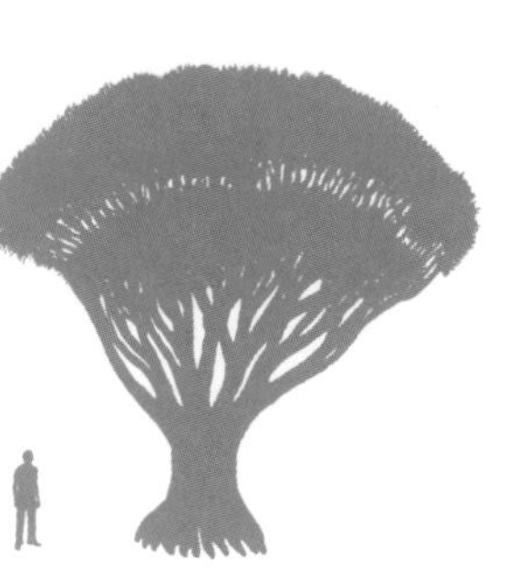

Height: up to 40 ft (12 m)
Type: evergreen semisucculent swordleaf
Occurrence: Canary Islands; Madeira; Cape Verde Islands; western Morocco
Habitat: semiarid scrubland on rocky hillsides and gullies
Division: Magnoliophyta: monocotyledons **Family:** Ruscaceae

The name Dragon Tree originates not from this tree's appearance but from a dark red, crystalline resin that exudes from any wound on its trunk or branches. This was called "dragon's blood" and was once valued for its medicinal properties as well as being used in varnishes. However, the major sources of dragon's blood resin were a tropical Asian palm and other *Dracaena* species from northeast Africa. In fact, closely related tree dracaenas are found in various parts of tropical Africa and Asia. Long-lived and slow-growing, it may take 20 years before the first panicle of flowers appears at the stem apex. Lateral branches only emerge after flowering; each branch may grow for another 10 years before it flowers.

The weathered lower branches of an old Dragon Tree show darker spots of dried resin. The trunk has bands of cells that can divide to increase its diameter, but no true wood.

Durian

Durio zibethinus

Height: up to 100 ft (30 m)
Type: evergreen broadleaf
Occurrence: western Malaysia and Indonesia; exact origin uncertain due to spread by humans
Habitat: tropical rain forest along river valleys, in regions with more than 60 in (1520 mm) annual rainfall
Division: Magnoliophyta: dicotyledons **Family:** Malvaceae

The Durian had a notorious reputation in European travelers' tales from Southeast Asia due to the fruit's reportedly disgusting smell, which had to be overcome before it could be tasted. But not all Durian fruit is foul-smelling; it depends on both the variety and stage of ripening. At its worst, it is like a combination of decaying soap in drains and an odorous European cheese. The pale yellowish pulp enclosing each large seed has a custard-like consistency and a cloyingly sweet, richly complex flavor. In some Sumatran markets, enthusiasts will stand around piles of fruit for hours testing and discussing the relative qualities of each fruit. Durian can only be grown in the lowland tropics.

The fruit has a thick, woody shell with strong, sharp prickles, which probably evolved as protection against orangutans and bears. The sweet-smelling flowers are pollinated by bats.

River Red Gum

Eucalyptus camaldulensis

Height: up to 140 ft (43 m)
Type: evergreen broadleaf
Occurrence: Australia: widespread except in Tasmania
Habitat: riverbanks, seasonally waterlogged plains, and channels of occasionally flowing streams
Division: Magnoliophyta: dicotyledons **Family:** Myrtaceae

This is one of the most important of the 800-plus species of *Eucalyptus* (all but a handful restricted to Australia) on two accounts. First, it has a wider range across the length and breadth of Australia than any other eucalypt. Second, it is the most widely grown plantation tree in the world's tropics and subtropics, at least for nonedible purposes. Fast-growing and with no large limbs during its young stages, it yields good-quality firewood as well as strong poles for construction, fencing and scaffolding. In some countries, such as Brazil, large quantities of the wood are burned for electricity generation. There is a large pool of genetic diversity within the species and some strains are very tolerant of drought, heat and poor soils.

River Red Gums are best known in Australia in the form of huge, old trees with massive, spreading limbs, in contrast to the spindly plantation trees seen in other countries.

Australian Mountain Ash

Eucalyptus regnans

Height: up to 320 ft (98 m)
Type: evergreen broadleaf
Occurrence: Tasmania and southern Victoria, Australia
Habitat: tall eucalypt forest on high-rainfall foothills and lower mountain areas
Division: Magnoliophyta: dicotyledons **Family:** Myrtaceae

Not only is this the tallest of all the *Eucalyptus* species and all Australian trees, it is the tallest nonconiferous tree in the world. The tallest living examples are exceeded only by the North American Coast Redwood (*Sequoia sempervirens*) and Douglas Fir (*Pseudotsuga menziesii*). But it is said that some Australian Mountain Ash felled more than 100 years ago were much taller—some reportedly grew to more than 400 ft (122 m)—which would make them taller than any living redwood. (Of course, the same may be said about the redwoods.) The name Mountain Ash comes from the tree's pale, straight-grained timber, which resembles that of the ashes (*Fraxinus* genus), not from any relationship to the European mountain ashes (*Sorbus* genus).

After wildfire, the Australian Mountain Ash grows quickly from seed in even-aged stands. Its straight trunk sheds branches cleanly. The lowest limb may be 100 ft (30 m) from the ground.

Candelabra Tree

Euphorbia ingens

Height: up to 36 ft (11 m)
Type: leafless succulent
Occurrence: tropical and subtropical eastern Africa from Kenya to South Africa
Habitat: semiarid scrubland and savanna on deep, sandy soil or rocky hills
Division: Magnoliophyta: dicotyledons **Family:** Euphorbiaceae

With about 2000 species, the *Euphorbia* genus is possibly the largest and most diverse genus of the flowering plants. It is found in most parts of the world and ranges from tiny, prostrate annuals to quite large trees. Common features are a thick, milky, acrid sap and the arrangement of the small, petalless flowers in a unique fleshy structure called a cyathium. About 850 species have evolved the succulent habit, leafless or almost so, with thick, often prickly stems. Most of these are African, ranging from golf ball-size (and shaped) plants to scrambling climbers and trees. In southern Africa alone there are about 20 species of tree euphorbias, reaching heights of up to 60 ft (18 m).

Euphorbia ingens is one of the largest tree euphorbias. The four to five angles of the stems are armed with rows of short prickles. The tree develops a candelabra-like form with age.

European Beech

Fagus sylvatica

Height: up to 120 ft (37 m)
Type: winter-deciduous broadleaf
Occurrence: throughout Europe except Britain; western Turkey
Habitat: temperate deciduous forest, mostly in valleys or on low hills, in well-drained soils
Division: Magnoliophyta: dicotyledons
Family: Fagaceae

Although this beech is one of England's best-loved trees, the evidence is strong that it was introduced there by Stone Age people, who would have used the nuts as food. It is one of about 10 *Fagus* species scattered across the northern hemisphere in Europe, Asia and America. They are a tight-knit group, differing only slightly in botanical characteristics, although of varying size and tree shape. The European Beech is fast-growing and not long-lived compared to the oaks (*Quercus* genus), but can develop a trunk over 6½ ft (2 m) in diameter. It yields a fine-grained timber that is easy to work, making it very suitable for small items such as handles and mallets. It is renowned for bentwood chairs.

Beech trees are admired for their sinuous trunks with smooth, gray bark and the thin, glossy leaves that shimmer gracefully, even in gentle breezes, and take on russet tones in fall.

Banyan

Ficus benghalensis

Height: up to 80 ft (24 m)
Type: evergreen broadleaf
Occurrence: India; Pakistan; Nepal; at low altitudes
Habitat: tropical rain forest and vine thickets, on deep, fertile soils
Division: Magnoliophyta: dicotyledons **Family:** Moraceae

The Banyan is remarkable for the way it sends its aerial roots down from along its thick, spreading limbs, then thickens these roots to form additional trunks. Over time, a single tree can grow into a forest of dozens or even hundreds of trunks supporting a single canopy of branches and foliage. Some famous specimens in India and Sri Lanka cover several acres and are hundreds of years old. It is true that a good proportion of the world's 800 or more fig (*Ficus*) species produce abundant aerial roots (see following entry), but few others behave in quite the same way as the Banyan. Its thick foliage provides good shade, under which villagers have traditionally gathered to trade and gossip. These trees have great cultural significance.

The Banyan's main trunk is heavily buttressed and larger than the secondary trunks formed from aerial roots. On some old trees it eventually rots away, leaving the many younger trunks.

Strangler Fig

Ficus virens

Height: up to 120 ft (37 m)
Type: dry-season-deciduous broadleaf
Occurrence: tropical and subtropical Asia; India; Japan; eastern Australia; Fiji
Habitat: lowland tropical and subtropical rain forest and vine thickets, usually in valleys and ravines
Division: Magnoliophyta: dicotyledons **Family:** Moraceae

This is just one of hundreds of *Ficus* species with the "strangler" habit of growth. It is one of the more wide-ranging of these figs and is also a very vigorous grower. A Strangler Fig starts life as a bird-deposited seed in the fork of a tree, needing light, humidity and air to germinate. The seedling soon sends down aerial roots that reach the ground and begin to thicken, fusing where they cross each other to form a network around the host tree's trunk. As this network grows and strengthens, it constricts the host's sap stream, eventually causing that tree's death. The advantage to the fig is that it gains early access to sunlight by starting high in a mature tree's crown.

This cascade of aerial roots comes from a large Strangler Fig that enveloped a host tree leaning at an angle. Every fig has a different, intricate network of aerial roots and buttresses.

Alerce

Fitzroya cupressoides

Height: up to 230 ft (70 m)
Type: evergreen conifer
Occurrence: southern Chile and Argentina, from 40° to 43° south latitude
Habitat: coniferous "rain forest" at low altitudes in regions with very high annual rainfall
Division: Pinophyta **Family:** Cupressaceae

This majestic conifer is said to grow taller than any other South American tree, though taller trees may lurk in the vast Amazon forests. Less in doubt is that it reaches a greater age than any other tree from that continent, with one tree dated by growth rings at more than 3600 years old. Spanish settlers in Chile called it *alerce,* Spanish for "larch," for its coniferous growth form and soft, durable timber. It was not until 1851 that it was recognized as a unique genus. It was named *Fitzroya* in honor of Captain Fitzroy of the *Beagle,* companion to Darwin, who reported seeing trees more than 30 ft (9 m) in diameter in Chile in 1834. The Alerce forests have since been vastly reduced by logging.

Alerce grows in moss-draped forests where rainfall may exceed 120 in (3050 mm) through the year. The climate can be mild to cold, but temperatures seldom fall below freezing.

Boojum Tree

Fouquieria columnaris

Height: up to 60 ft (18 m)
Type: succulent dry-season-deciduous
Occurrence: central Baja California and nearby coastal Sonora, Mexico
Habitat: very hot, arid scrubland on stony hills
Division: Magnoliophyta: dicotyledons **Family:** Fouquieriaceae

In its native northwest Mexico, this bizarre plant is known as *cirio* (Spanish for "wax candle"). But English speakers from north of the border gave it the whimsical name Boojum Tree, from the mysterious beast, the Boojum, in Lewis Carroll's *The Hunting of the Snark*. Its unbranched, thorny growth form is very striking, but a closer look at flowers and foliage reveals its relationship to other flowering plants. The 11 *Fouquieria* species are all confined to this region of North America, and are the only members of the Fouquieriaceae family. But that is closely allied to the much larger Polemoniaceae, or phlox, family. Other *Fouquieria* species are more shrubby, including the red-flowered Ocotillo (*F. splendens*), which extends into California and Arizona.

Boojum Trees create a surreal landscape in the Catavina region of Baja California Peninsula. Side branches are abortive, becoming thornlike. Sprays of cream flowers appear in summer.

Franklin Tree

Franklinia alatamaha

Height: up to 20 ft (6 m)
Type: winter-deciduous broadleaf
Occurrence: Georgia, USA: extinct in the wild
Habitat: forest or scrubland along boggy streams and among old dunes on coastal plains
Division: Magnoliophyta: dicotyledons **Family:** Theaceae

This small tree may be the most famous example of an ornamental plant that is extinct in the wild but preserved by cultivation. Early plant explorers John and William Bartram found it in 1765 in the delta region of Georgia's Altamaha River, then a sandy wilderness. They later returned to obtain seeds and grew it in their Philadelphia garden. In 1785 their cousin, self-taught botanist Humphrey Marshall, named it as the new genus *Franklinia,* honoring his friend and Philadelphia's most famous citizen, Benjamin Franklin. The last sighting of wild plants was in 1803. It is thought that the spread of cotton planting and associated plant diseases may have caused its extinction. But the species thrives still in many gardens.

The Franklin Tree flower shows a family resemblance to camellias. The tree blooms at the end of summer and into fall, when the leaves begin to turn wine red with orange flushes.

Ginkgo

Ginkgo biloba

Height: up to 120 ft (37 m)
Type: deciduous gymnosperm
Occurrence: central, eastern China: possibly extinct in the wild
Habitat: temperate deciduous forest on steep mountainsides
Division: Ginkgophyta
Family: Ginkgoaceae

The popular story about the Ginkgo is that it was extinct in the wild but had been preserved for 1000 years or more by being planted around a few temples and monasteries. But as China's plant life becomes better known, it seems that there may still be wild stands of this remarkable tree. Moreover, it has been widely cultivated for many centuries, in Japan and Korea as well as China, for its edible seeds as well as for its ornamental qualities. But what's more amazing is that the genus has survived with only minor changes since the Jurassic period, and its family appears even farther back in the Permian period, almost 270 million years ago. Fossils show that Ginkgo was common in all continents up until the early Tertiary period.

Ginkgo is easy to grow and long-lived. The leaves turn golden yellow in fall. Female trees bear large, yellow-skinned, naked seeds with edible kernels; the outer "flesh" of fallen seeds stinks.

Bago

Gnetum gnemon

Height: up to 60 ft (18 m)
Type: evergreen gymnosperm
Occurrence: Southeast Asia; western Pacific from eastern India to Fiji
Habitat: tropical rain forest, usually close to streams
Division: Gnetophyta **Family:** Gnetaceae

What is so interesting about this small, rather nondescript rain forest tree is where its genus, *Gnetum*, fits into plant evolution. *Gnetum* lacks any fossil record. But molecular and other evidence shows that it belongs with two other genera of totally different appearance: the bizarre *Welwitschia* of the Namib Desert, and the northern hemisphere joint firs, the *Ephedra* genus, which is the source of ephedrine. These three make up the Gnetophyta, which is most closely related to the Pinophyta (conifers). The gnetums are true gymnosperms with naked seeds. They have catkins of male and female organs, sometimes on different trees. The large seed kernels of some species are used for food. There are 28 species in all, but the Bago is the only one that forms a tree.

This close-up of a Bago catkin shows the male (cream) and female (green) organs. Although unrelated to the flowering plants, the Bago's broad leaves resemble their leaves.

Silky Oak

Grevillea robusta

Height: up to 120 ft (37 m)
Type: semideciduous broadleaf
Occurrence: southeastern Queensland and northeastern New South Wales, Australia
Habitat: drier types of subtropical rain forest, on fertile soils
Division: Magnoliophyta: dicotyledons **Family:** Proteaceae

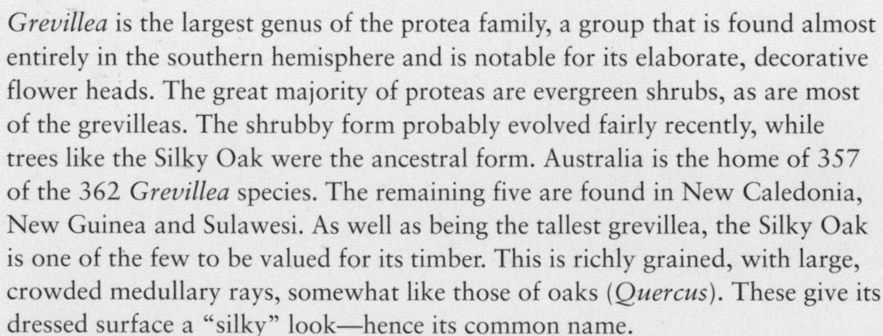

Grevillea is the largest genus of the protea family, a group that is found almost entirely in the southern hemisphere and is notable for its elaborate, decorative flower heads. The great majority of proteas are evergreen shrubs, as are most of the grevilleas. The shrubby form probably evolved fairly recently, while trees like the Silky Oak were the ancestral form. Australia is the home of 357 of the 362 *Grevillea* species. The remaining five are found in New Caledonia, New Guinea and Sulawesi. As well as being the tallest grevillea, the Silky Oak is one of the few to be valued for its timber. This is richly grained, with large, crowded medullary rays, somewhat like those of oaks (*Quercus*). These give its dressed surface a "silky" look—hence its common name.

The flowers open after the Silky Oak sheds all or most of its leaves in late spring, offering rich nectar rewards to birds and bees. The tree is widely cultivated in warm climates.

Rubber Tree

Hevea brasiliensis

Height: up to 100 ft (30 m)
Type: evergreen broadleaf
Occurrence: Amazon Basin: mainly Brazil but also bordering countries
Habitat: midstory of tropical rain forest
Division: Magnoliophyta: dicotyledons **Family:** Euphorbiaceae

The demand for rubber began in the 19th century. It was very useful in many developing technologies, especially once vulcanization was discovered. By the midcentury the best rubber was coming from the Brazilian state of Pará, gathered from the Rubber Tree by an army of rubber tappers. In 1875 the British smuggled seeds out of Brazil and established seedlings in Sri Lanka and Singapore. Plantations on the Malay Peninsula were soon producing rubber more cheaply than Brazil. This tree was once known as Para Rubber to distinguish it from India Rubber (*Ficus elastica*) and other once-promising rubber-yielding trees. It belongs to the huge and diverse euphorbia family, many members of which exude a thick, milky latex from their stems when cut.

Rubber tapping in a Southeast Asian plantation—each morning, a new spiral groove is cut in the bark to intersect latex channels that run in opposite spirals. A cup collects the latex.

Jacaranda

Jacaranda mimosifolia

Height: up to 80 ft (24 m)
Type: dry-season-deciduous broadleaf
Occurrence: southern Brazil; Bolivia; Paraguay; northern Argentina
Habitat: tropical and subtropical deciduous forest on plateaus
Division: Magnoliophyta: dicotyledons
Family: Bignoniaceae

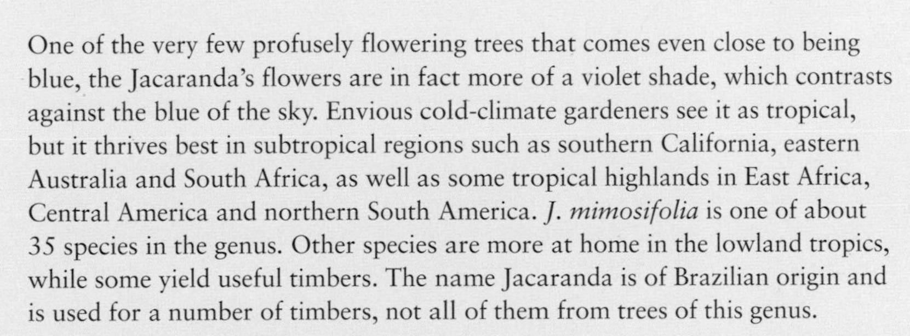

One of the very few profusely flowering trees that comes even close to being blue, the Jacaranda's flowers are in fact more of a violet shade, which contrasts against the blue of the sky. Envious cold-climate gardeners see it as tropical, but it thrives best in subtropical regions such as southern California, eastern Australia and South Africa, as well as some tropical highlands in East Africa, Central America and northern South America. *J. mimosifolia* is one of about 35 species in the genus. Other species are more at home in the lowland tropics, while some yield useful timbers. The name Jacaranda is of Brazilian origin and is used for a number of timbers, not all of them from trees of this genus.

Jacaranda's ferny, bipinnate leaves drop in late winter. Flowering starts in late spring just before the new leaves appear. Flower shape, color and markings ensure pollination by bees.

Black Walnut

Juglans nigra

Height: up to 140 ft (43 m)
Type: winter-deciduous broadleaf
Occurrence: eastern half of USA
Habitat: temperate deciduous forest, in deep, alluvial soils on riverbanks and bottomlands
Division: Magnoliophyta: dicotyledons
Family: Juglandaceae

The Black Walnut is a multipurpose tree. The nuts, though slightly smaller than those of the Eurasian Walnut (*J. regia*), are equally good for eating. The very hard shells have a surprising range of industrial uses when ground up, from metal-cleaning abrasives to a filler for dynamite. The fine-grained lumber is renowned for its use in furniture, paneling and gunstocks. Fruit husks can yield a yellowy brown dye. Black Walnut grows well in plantations on deep, fertile soils. All parts of the tree contain the compound juglone. This is released into the soil when the tree's litter breaks down, where it kills or stunts competing plants. Gardeners know that few plants thrive under a walnut tree.

The massive crown of a mature Black Walnut—the pinnate leaves are beginning to turn their deep gold fall color. The prolific, hard fruits can be a hazard when they rain down.

Utah Juniper

Juniperus osteosperma

Height: up to 40 ft (12 m)
Type: evergreen conifer
Occurrence: western USA from California northeast to Montana and southeast to New Mexico
Habitat: semiarid piñon–juniper woodland and scrubland at 3000–8000 ft (910–2440 m)
Division: Pinophyta **Family:** Cupressaceae

This is typical of a group of juniper species found over vast areas of range country in western North America. It occurs in large numbers on the upland plateaus, most extensively in Utah. Its co-dominant trees are the compact pinyon pines (*Pinus edulis* and *P. monophylla*) and several small, evergreen oaks. There are about 60 *Juniperus* species scattered through the northern hemisphere, most being very cold-hardy trees or shrubs. They are true conifers. Their seed cones are quite uncone-like, however, resembling small berries because their soft, fleshy scales are fused together. In most American species the foliage is cypress-like, the leaves being reduced to tiny scales.

A gnarled Utah Juniper stands on a cliff rim at Utah National Monument. The short, single trunk is characteristic. Hopi and Navajo people had many uses for junipers, some medicinal.

Huon Pine

Lagarostrobos franklinii

Height: up to 100 ft (30 m)
Type: evergreen conifer
Occurrence: southwestern Tasmania, Australia
Habitat: riverbanks and swampy areas in cool temperate rain forest dominated by Southern Beeches (*Nothofagus*)
Division: Pinophyta **Family:** Podocarpaceae

This, the only species of *Lagarostrobos*, a genus unique to Tasmania, is one of several conifers endemic to that island. All have strong affinities to conifers of other cool southern hemisphere lands, especially New Zealand and Chile. But the Huon Pine is not a pine (*Pinus*) at all. It is one of the podocarps, an ancient family of conifers that ranges from high southern latitudes to as far north as Mexico and central China, including tropical rain forests. The Huon Pine is renowned for the great age of some specimens—more than 2500 years old. Also, its timber is highly valued for boat-building, fine cabinetwork and woodcraft. Nowadays, it is protected. The only supplies come from areas inundated by dams, with salvaged timber stockpiled by the government.

These moss-draped Huon Pines may be more than 1000 years old, their trunks increasing in diameter by as little as 1/25 in (1 mm) a year. The wood is light, strong and durable.

Tulip Tree

Liriodendron tulipifera

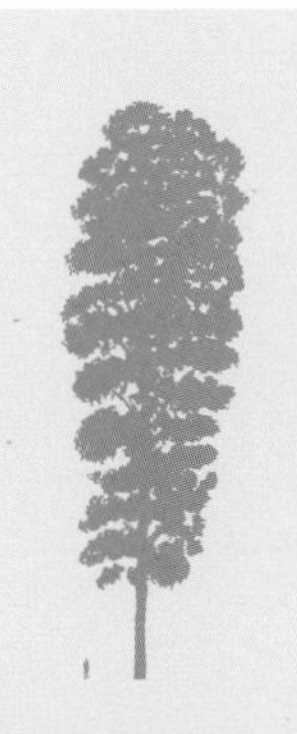

Height: up to 180 ft (55 m)
Type: winter-deciduous broadleaf
Occurrence: eastern USA from the Great Lakes to northern Florida
Habitat: temperate deciduous forest on sheltered hill slopes and hollows, mostly at low altitudes
Division: Magnoliophyta: dicotyledons **Family:** Magnoliaceae

This is one of the USA's most useful timber trees, valued for its rapid growth and general-purpose lumber, which is sold under the name "Yellow Poplar," although the tree is unrelated to the poplars (*Populus*). In fact the *Liriodendron* genus is the closest relative of the large and diverse *Magnolia* genus. It is an example of a split Chinese–eastern North American distribution, as the only other species occurs in China. Fossils show the genus was more widespread in the northern hemisphere during the Tertiary period. The Tulip Tree's botanical name means "tulip-bearing lily tree," and its flowers do have a slightly tulip-like appearance. It differs from magnolias in having papery, winged seeds.

The Tulip Tree makes a fine ornamental for parks and large gardens. The 2–3-in (5–7.5-cm) green and orange summer flowers are followed by butter yellow foliage in fall.

Southern Magnolia

Magnolia grandiflora

Height: up to 120 ft (37 m)
Type: evergreen broadleaf
Occurrence: southeastern USA from eastern Texas to northern Florida
Habitat: temperate deciduous forest and pine forest in rich, alluvial soils on coastal lowlands
Division: Magnoliophyta: dicotyledons **Family:** Magnoliaceae

Alphabetical coincidence has the Southern Magnolia following *Liriodendron,* the only other genus in its family. Molecular studies have shown other genera in the family must now be merged into an expanded *Magnolia* genus with more than 220 species. It occurs in the Americas and Asia, but not in Europe, Africa or Australia. The family is ancient, one of the "primitive" families that diverged from ancestral flowering plants even before the major branch divided into monocotyledons and dicotyledons. This tree, also called the Bull Bay, is among the largest temperate magnolias. It is used almost entirely as an ornamental, popular in warm-temperate parks and gardens around the world.

The Southern Magnolia has some of the largest flowers in the *Magnolia* genus. Borne through summer and into fall among large, glossy leaves, they are spicily fragrant.

Tropical Paperbark

Melaleuca leucadendra

Height: up to 120 ft (37 m)
Type: evergreen broadleaf
Occurrence: tropical Australia; New Guinea; the Moluccas
Habitat: riparian forest on riverbanks, channels, dunes and swamps
Division: Magnoliophyta: dicotyledons **Family:** Myrtaceae

The *Melaleuca* genus comprises more than 220 species, all but 10 being endemic to Australia, where many are only shrubs. They usually have tough, harsh-textured leaves and bottlebrush-like spikes of flowers with numerous showy white, pink or red stamens. These are basally fused into five bundles, a feature that defines the genus. The Tropical Paperbark is one of the larger *Melaleuca* tree species that have thick, whitish bark, which separates into papery layers. Some species extend beyond Australia to New Caledonia, New Guinea, Indonesia and one (*M. cajuputi*) even to mainland Southeast Asia. Another Australian member of this group, *M. quinquenervia,* has been widely planted for ornament but is now the worst weed tree of the Florida Everglades.

On riverbanks and floodplains in northern Australia, the Tropical Paperbark becomes a majestic tree with pendulous branchlets. White flower spikes precede tiny seed capsules.

Dawn Redwood

Metasequoia glyptostroboides

Height: up to 160 ft (49 m)
Type: deciduous conifer
Occurrence: Sichuan–Hubei–Hunan border region, central China
Habitat: fringing forest along riverbanks, in deep, alluvial soil
Division: Pinophyta **Family:** Cupressaceae

This rare Chinese tree is often grouped with the Ginkgo (*Ginkgo biloba*) as one of the most ancient "living fossils." However, its known fossils are only about half as old as the Ginkgo's, no more ancient than many conifers and even some flowering plants. The genus was first known to science as fossils from Japan, Siberia and North America, and its name was published by a Japanese paleobotanist in 1941 (*Metasequoia* means "beyond Sequoia"). In 1944 the botanist Zhan Wang was shown strange trees in a village in Sichuan. These were identified as the *Metasequoia* genus and named *M. glyptostroboides* in 1948. Harvard University's Arnold Arboretum funded an expedition to collect seeds in 1948, resulting in Dawn Redwood's cultivation around the world.

Population pressure endangers it in the wild, but the Dawn Redwood is popular in gardens for its conical form and ferny foliage. It sheds its two-rowed leaves on short branchlets in fall.

Pohutukawa

Metrosideros excelsa

Height: up to 60 ft (18 m)
Type: evergreen broadleaf
Occurrence: northern half of North Island, New Zealand
Habitat: evergreen coastal forest and scrubland, on rocky bluffs and beach dunes
Division: Magnoliophyta: dicotyledons **Family:** Myrtaceae

The Pohutukawa is New Zealand's most colorful flowering tree. Its canopy bursts into a sheet of brilliant red blossom in early summer, which earns it the alternative name of New Zealand Christmas Tree in the southern hemisphere. Stretches of steep coastline, such as in the Bay of Islands, are famed for the massed flowering of wind-shorn Pohutukawa. As with the *Eucalyptus* and *Callistemon* genera—which are also members of the myrtle family, though not close allies of *Metrosideros*—the Pohutukawa's vivid display comes from its flowers' long stamens, which attract birds and insects as pollinators. The genus consists of about 50 species, mainly scattered throughout Pacific Islands as far east as Hawaii and the Marquesas. The fine seeds are carried by storm winds.

Pohutukawa is reduced to a shrub by exposure on sea cliffs. Its toughness in extreme coastal conditions has made it a favorite tree for seafront plantings in frost-free, temperate regions.

Horseradish Tree

Moringa oleifera

Height: up to 30 ft (9 m)
Type: dry-season-deciduous broadleaf
Occurrence: India: origin uncertain due to spread by humans, but possibly Himalayan foothills
Habitat: tropical deciduous woodland and vine thickets, on dry hill slopes
Division: Magnoliophyta: dicotyledons **Family:** Moringaceae

This is another "miracle tree," hailed by some as the answer to the needs of the developing world in the drier tropics. And it does have many virtues. It can survive in drought and poor soil, and yields edible, nutritious shoots, flowers and fruit, as well as a seed oil with many uses. *Moringa* has 13 species of small trees and shrubs that occur in the drier parts of east Africa and southwest Asia. It is the only genus of the Moringaceae family, which is part of the larger Brassicales order. Most Brassicales members contain glucosinolates, sulfur-based compounds that release the "mustard oils" that give cabbages and their relatives, such as horseradish and mustard, their distinctive sharp flavors.

The Horseradish Tree has a short trunk, lanky branches and large leaves. Multipetaled, cream flowers appear with the new leaves before the long, edible fruits, sold as "drumsticks" in India.

Nutmeg Tree

Myristica fragrans

Height: up to 60 ft (18 m)
Type: evergreen broadleaf
Occurrence: eastern Indonesia: origin possibly Banda Islands in the Moluccas
Habitat: tropical rain forest
Division: Magnoliophyta: dicotyledons **Family:** Myristicaceae

The product of this tree was one of the most precious spices to be traded westward through Asia and on to Europe since ancient times. Its origin in these tiny specks of far-eastern islands was shrouded in mystery until the Portuguese began long-distance sea travel from Europe. Then followed centuries of national rivalry to corner the lucrative trade. The spice comes from the Nutmeg Tree's large, hard seed, which is contained in a fleshy capsule that splits into two halves. The seed is covered by a bright red aril (outgrowth of seed stalk), which yields the tree's other spice, mace. Harmless in small pinches, nutmeg is hallucinogenic in doses of only a quarter of an ounce (a few grams) and acutely toxic to the liver in slightly higher doses.

Fruits of the Nutmeg Tree split open to show the red seed arils. *Myristica* is one of several genera in the family, which diverged from other flowering plants very early in its evolution.

Balsa

Ochroma pyramidale

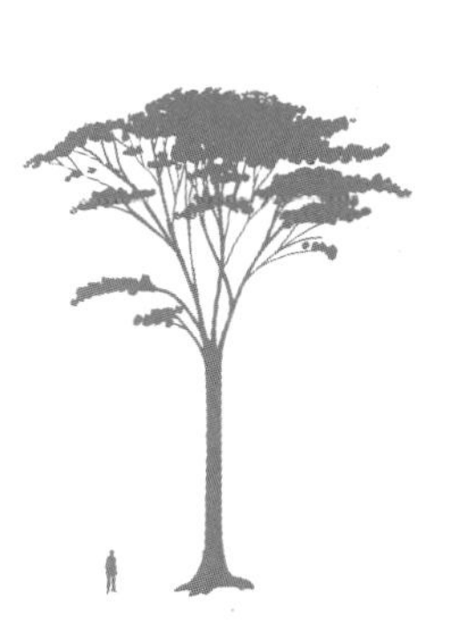

Height: up to 100 ft (30 m)
Type: evergreen broadleaf
Occurrence: southern Mexico; Central America; Andean South America to Bolivia
Habitat: secondary rain forest colonizing disturbed sites, such as landslips and hurricane-created clearings, often on poor soils
Division: Magnoliophyta: dicotyledons **Family:** Malvaceae

The Balsa is far better known for its wood than as a tree, due to its lightness. Specific gravity after drying can be as low as 0.1, or one-tenth the density of water. Balsa wood has many uses where lightness and strength are required. However, it has been replaced by synthetics in many former applications, such as surfboards and lifebuoy cores. Its major producer is Ecuador, where logs are floated down rivers in the form of rafts (*balsa* is Spanish for "raft"). These days, adventure tourism companies offer rides on these rafts, adding value to the logging. The most famous balsa raft was Thor Heyerdahl's *Kon-Tiki,* a 1947 reconstruction of large, oceangoing rafts of prehistoric Ecuadorians.

The large, cream flower of the Balsa starts to open in the evening. It is pollinated by bats that feed on its copious nectar. The leaves are almost circular, and up to 1 ft (30 cm) across.

Olive

Olea europaea

Height: up to 40 ft (12 m)
Type: evergreen broadleaf
Occurrence: origin uncertain, possibly North Africa or southwest Asia
Habitat: evergreen woodland and scrubland, in shallow soils on rocky slopes
Division: Magnoliophyta: dicotyledons **Family:** Oleaceae

Mediterranean food is based on three plants: the fig, the olive and the grape. All have been cultivated for millennia and modified so much by humans that their wild ancestors are not readily identifiable. Olive pits have been found in 6000-year-old excavated dwellings in the Palestine region. The *Olea* genus has about 30 species, mostly broader-leaved trees in Asia and Africa. The closest wild relatives of O. *europaea* have much smaller fruit and seeds. In ancient Greece and Rome, olive oil was a major trade item, transported in large amphorae carried on racks in ships. The health-preserving role of olive oil in the Mediterranean diet has been scientifically proven.

An old Olive tree grows in the harsh, rocky terrain found around the Mediterranean. However, most commercial orchards are planted on deeper soils, often with irrigation.

Baja Elephant Tree

Pachycormus discolor

Height: up to 30 ft (9 m)
Type: dry-season-deciduous broadleaf
Occurrence: central Baja California, northwest Mexico
Habitat: hot, arid scrubland and steppes, on stony hillsides and plateaus
Division: Magnoliophyta: dicotyledons **Family:** Anacardiaceae

This is a prime example of a pachycaul (literally "thick-stemmed") tree, a growth form that several plant families have evolved as an adaptation to arid environments. Pachycauls include the Baobab (*Adansonia digitata*) and Boojum Tree (*Fouquieria columnaris*). Indeed, the latter is often seen growing next to the Baja Elephant Tree. A typical pachycaul has a greatly swollen lower trunk, but its branches, leaves and flowers are characteristic of the family it belongs to. The leaves are usually shed in dry seasons. Many shrubs and vines also have a pachycaul habit. Such plants have a devoted following among one group of succulent plant enthusiasts, who call them "caudiciforms." The genus contains this single species, and is allied to the sumacs (*Rhus* genus).

The Baja Elephant Tree's trunk is up to 3 ft (90 cm) in diameter, but it is usually under 20 ft (6 m) tall. The bark peels in thin, translucent layers. It has green photosynthetic inner bark.

Beach Pandanus

Pandanus tectorius

Height: up to 30 ft (9 m)
Type: evergreen swordleaf
Occurrence: tropical coasts of Pacific and Indian Oceans from Hawaii westward to Sri Lanka
Habitat: seashore, at backs of ocean beaches and on rocky slopes and headlands exposed to salt spray
Division: Magnoliophyta: monocotyledons **Family:** Pandanaceae

Pandanus is a genus of over 700 species, all but a few of which grow in limited parts of the Old World tropics, mostly the Malay Archipelago and Madagascar. But this species is found almost throughout the genus's range. *Pandanus* has very distinctive features. Long, channeled leaves with prickly edges and rows of prickles on the underside are arranged in clear spirals on the stems. Many species have stilt roots at the main stem's base. Male and female flowers are on separate plants, the female's in round heads that develop into large fruiting heads with a complex construction. Each stem branches after flowering. The branching is repeated at intervals, producing the characteristic forking pattern.

An old Beach Pandanus extends its branches over the sea in Hawaii, the plant's most northeastern outpost. Its Polynesian names are *hala* or *fara*.

Princess Tree

Paulownia tomentosa

Height: up to 60 ft (18 m)
Type: winter-deciduous broadleaf
Occurrence: northern and eastern China
Habitat: temperate deciduous forest and woodland, on deep, valley soils
Division: Magnoliophyta: dicotyledons **Family:** Paulowniaceae

Paulownia is a remarkable genus of six species, mostly unique to China. The Princess Tree was the first one known in the West. It was named *Paulownia imperialis* in 1835 after the Russian Princess Anna Paulowna (Pavlovna), but the existence of an older name, *Bignonia tomentosa,* required that the species be changed to *tomentosa*. All paulownias have large, simple leaves in opposite pairs. On vigorous saplings the leaves can be huge, up to 18 in (46 cm) wide. The plants grow very rapidly through summer. This fast growth and their very tough but light wood have long been valued in China, and in recent years some species have been promoted as plantation trees in countries such as the USA. However, there has been concern at their increasing weediness.

The Princess Tree has the most colorful flowers of all paulownias, its crown a mass of pale violet in midspring. Flowers open from rust brown calyxes that develop through late winter.

Date Palm

Phoenix dactylifera

Height: up to 80 ft (24 m)
Type: solitary or clustered feather palm
Occurrence: origin uncertain due to spread by humans: possibly Egypt–Sudan region
Habitat: river floodplains and channels of nonpermanent streams
Division: Magnoliophyta: monocotyledons **Family:** Arecaceae

The Date Palm has the greatest significance to the food and culture of the hot, dry lands of North Africa and the Middle East. It bears fruit that is delicious when fresh but is also very usable in dried form. Indeed, dried dates are a concentrated energy source that have enabled nomadic herders to survive their desert travels. Date Palms tolerate fierce heat but must have permanent soil moisture, and so are planted along rivers such as the Nile or in oases and desert wadis where aquifers surface. Hand-pollination of female flower panicles with pollen from male flowers, which are found on separate trees, is commonly practiced to ensure fruiting.

To produce dates with good sugar content, this palm needs a long, hot summer and dry air. Southern California is one place outside their region of origin where they have succeeded.

Ombú

Phytolacca dioica

Height: up to 60 ft (18 m)
Type: evergreen broadleaf
Occurrence: northern Argentina; southern Brazil; Uruguay; Paraguay; Bolivia
Habitat: subtropical woodland and savanna, mostly on elevated plateaus
Division: Magnoliophyta: dicotyledons **Family:** Phytolaccaceae

The striking feature of a mature Ombú is the way its trunk base spreads into a vast cone, up to 20 ft (6 m) in diameter at soil level, even though the tree may only be 30 ft (9 m) tall. It has a pachycaul growth habit, like the Baja Elephant Tree (*Pachycormus discolor*). However, the Ombú may have evolved this less as a response to aridity than as a means of providing a stable support for its heavy crown of foliage. The *Phytolacca* genus consists chiefly of herbaceous plants and shrubs that may have an inherent inability to develop normal wood. Ombú trunks thicken with successive new layers of cambium tissue, creating weak, fleshy stem tissues that are partly supported by internal sap pressure.

Ombú leaves, flowers and fruit are similar to those of other *Phytolacca* species, such as Common Pokeweed (*P. americana*). Ombú has male and female flowers on separate trees.

Sitka Spruce

Picea sitchensis

Height: up to 320 ft (98 m)
Type: evergreen conifer
Occurrence: west coast of North America, from far northwestern California to southwestern Alaska
Habitat: coniferous "rain forest" along estuary shores, in coastal valleys and on foothills up to 2000 ft (610 m)
Division: Pinophyta **Family:** Pinaceae

The Sitka Spruce is the tallest and most vigorous of the 33 *Picea* species. This genus occurs in cooler parts of the northern hemisphere, and includes the northernmost conifers in Eurasia and North America, which are abundant in the taiga. The Sitka Spruce, in contrast, is restricted to the more sheltered coastal "rain forests" where winter temperatures seldom fall below 0°F (–18°C), even in Alaska, and annual rainfall can be as much as 150 in (3810 mm). Its timber has outstanding qualities, combining stiffness with light weight. As a result, Sitka Spruce forests have been heavily exploited. Once used for aircraft frames, the timber is still prized for musical instruments and boat-building.

The Sitka Spruce grows in moss-draped forests, often in quite boggy ground with a thick layer of dead organic matter on the soil surface, and always within 100 miles (160 km) of the sea.

Bristlecone Pine

Pinus aristata

Height: up to 60 ft (18 m)
Type: evergreen conifer
Occurrence: Utah, Nevada and eastern California from Rockies to White Mountains, USA
Habitat: coniferous woodland on stony summits and slopes at 6500–11,800 ft (1980–3600 m), often bare of soil and understory
Division: Pinophyta **Family:** Pinaceae

The same pines that survive the harshest mountain environments are also the oldest living trees, with ages of up to 4700 years according to growth rings. Often, most of the tree is dead wood, with only narrow strips of live bark sustaining foliage growth. The botanical name *P. aristata* is used here in a broad sense to include these oldest trees, which since 1970 have usually been treated as a separate species under the name *P. longaeva,* though distinguishing features are slight. Even after death, Bristlecone Pine trunks may survive many hundreds of years. This allows cross-matching of their growth rings with living trees and the reconstruction of climate fluctuations going even farther back.

This ancient Bristlecone Pine may survive many more years yet. Seed production is prolific, and storage of seeds under stones by the Clark's nutcracker bird may aid germination.

Lacebark Pine

Pinus bungeana

Height: up to 100 ft (30 m)
Type: evergreen conifer
Occurrence: north-central China
Habitat: mixed coniferous and broadleaf forest, on rocky hills and mountains up to 6000 ft (1830 m)
Division: Pinophyta **Family:** Pinaceae

The Lacebark Pine is one of the most beautiful of all the 111 species of true pines, thanks to its elegant form and dappled bark. It is best known from the old planted trees found in the grounds of Beijing's Forbidden City and at other sites near Beijing. The most famous specimen is the Nine Dragon Pine at Jie Tai Temple, the trunk of which is branched into nine. It is said to be over 900 years old, and is believed to be the same tree from which Russian botanist Alexander Bunge collected a specimen in 1831. The name *Pinus bungeana* was published in Germany in 1847, based on his collection. The needles of the Lacebark Pine are short and glossy, deep green. The small cones have few leathery scales and large seeds, which are used as food in China.

On old Lacebark Pines, the bark pattern is overlaid with white, giving a dramatic silvery look to the trunk and limbs. On younger trees like this, it is dappled green, brown and gray.

Ponderosa Pine

Pinus ponderosa

Height: up to 230 ft (70 m)
Type: evergreen conifer
Occurrence: western half of USA east to Nebraska; southwestern Canada
Habitat: temperate coniferous forests, mainly on hills and plateaus, on moderately fertile soils
Division: Pinophyta **Family:** Pinaceae

This is the most widely distributed North American pine as well as the most important as a source of lumber. Its botanical epithet, "ponderosa," bestowed by explorer David Douglas, who discovered it in 1826, is Latin for "heavy," referring to the density of its wood compared with other pines. That epithet has since come into popular usage. The Ponderosa Pine is a majestic tree on better sites. It has a shaftlike trunk free of lower branches, a narrow crown with "powder-puff" rosettes of long needles, and orange brown bark in broad plates separated by gray fissures. Each scale on the cones has a sharp prickle on the back. It is a pine that tolerates forest fires if they are not too fierce.

Forests of Ponderosa Pine typically have a clear understory, lacking shrubs and smaller trees. They are habitat for mammals such as deer. The seeds provide food for squirrels and birds.

Monterey Pine

Pinus radiata

Height: up to 200 ft (61 m)
Type: evergreen conifer
Occurrence: central coastal California, USA; islands off northern Baja California, Mexico
Habitat: coniferous forest on coastal hills below 1000 ft (300 m) on mainland, up to 3600 ft (1100 m) on islands.
Division: Pinophyta **Family:** Pinaceae

This pine is of minor significance in its native California but of major importance as a plantation tree in other countries, most notably New Zealand, Australia, Chile and South Africa. In fact, conditions in New Zealand and Australia's southeast suit it so well that numerous trees there are much taller than the tallest known trees in California. No doubt they are helped by the absence of natural pests and diseases—the same reason Australia's eucalypts often grow better beyond its shores. The timber, produced in huge quantities from plantations and known as Radiata Pine, is pale and of medium density. It is used for furniture, house frames, chipboard and paper pulp.

The Monterey Pine has tangled, grass green needles. It is one of several Californian pines that hold their seed in closed cones for years, an adaptation to fire-prone habitats.

Eastern White Pine

Pinus strobus

Height: up to 200 ft (61 m)
Type: evergreen conifer
Occurrence: northeastern USA; southeastern Canada and USA from Newfoundland to Georgia; southern Mexico; Guatemala
Habitat: coniferous forest from sea level in north to 5000 ft (1520 m) in south; up to 8000 ft (2440 m) in Mexico
Division: Pinophyta **Family:** Pinaceae

It was this pine that largely formed the "interminable forest" so graphically depicted in James Fenimore Cooper's novel, *The Last of the Mohicans*. This was set in the mid-18th century, when, from the eastern seaboard, the pine forests clothing the Appalachian Mountains seemed vast, gloomy and impenetrable. Even then, they were being exploited for ships' spars and housing, and by the mid-19th century the old-growth forest was much depleted. The species is still abundant, however. It is a typical member of the "soft pines" subgenus. It has soft, whitish wood, blue green needles in clusters of five with white stomatic bands along each needle, and narrow, thin-scaled seed cones.

A stand of Eastern White Pine grows by a river. In some regions the pine's vigor has been reduced by white pine blister rust, often in combination with pollution such as acid rain.

Oriental Plane

Platanus orientalis

Height: up to 120 ft (37 m)
Type: winter-deciduous broadleaf
Occurrence: Himalayas west from Kashmir; Iran; Turkey; Balkans
Habitat: temperate deciduous forest and woodland on rocky slopes and in ravines
Division: Magnoliophyta: dicotyledons **Family:** Platanaceae

The planes (*Platanus*) are so familiar in streets and parks that it comes as a surprise to learn that they form a botanically unique family, traceable as fossils back to more than 110 million years ago, with no close relatives. Molecular evidence shows that the nearest related families are the seemingly very different protea family (Proteaceae) of the southern hemisphere, and the aquatic lotus family (Nelumbonaceae). There are ten *Platanus* species. Eight are native in Mexico and North America, where they are called sycamore or buttonwood. One little-known evergreen species comes from Laos and Vietnam. And there is the Oriental Pine of Eurasia. It and the American *P. occidentalis* are best known in their hybrid form, *P.* x *acerifolia*, the famous London Plane.

The majestic Oriental Plane has large, spreading limbs and mottled, flaking bark. Prized for summer shade, it was under a plane (still living) that Hippocrates taught his students on Kos.

Cottonwood

Populus deltoides

Height: up to 170 ft (52 m)
Type: winter-deciduous broadleaf
Occurrence: southern Canada; eastern and central USA; northern Mexico
Habitat: deciduous broadleaf forest on riverbanks, sandbars, floodplains and bottomlands, on deep, alluvial soils
Division: Magnoliophyta: dicotyledons **Family:** Salicaceae

The Cottonwood is the largest-growing poplar species, the most wide-ranging in North America, and the most important for timber. It is renowned for its rapid growth in ideal conditions—up to 100 ft (30 m) in 9 years has been recorded in a southern forest plantation. The whitish, low-density wood has many uses, from matchsticks, clothes-pegs and clogs to machinery cases and plywood. Cottonwood is a rather ambiguous name. It broadly refers to a number of American species in the black poplar and balsam poplar subgenera. *P. deltoides* is divided into three subspecies: Eastern Cottonwood, subsp. *deltoides;* Plains Cottonwood, subsp. *monilifera;* and Rio Grande Cottonwood, subsp. *wislizenii.*

These Cottonwoods in a valley bottom show the spreading form of the Plains Cottonwood, which extends the range of the species west to the North American Continental Divide.

Douglas Fir

Pseudotsuga menziesii

Height: up to 330 ft (100 m)
Type: evergreen conifer
Occurrence: western North America from central British Columbia southeast to Utah and south to central Mexico
Habitat: coniferous forest in high-rainfall coastal valleys and foothills; drier coniferous forest in mountains to 9500 ft (2900 m)
Division: Pinophyta **Family:** Pinaceae

The Douglas Fir is a remarkable tree. It is the second-tallest conifer after the Coast Redwood (*Sequoia sempervirens*) and the tallest tree in the Pinaceae family. It occurs in a greater latitude range than almost any other American conifer. Also, it is one of the world's most important timber trees. The *Pseudotsuga* genus is characterized by short needle leaves, like the spruces, but it has cone scales tipped by conspicuous three-lobed "bracts." The species *P. menziesii* can be divided into the very tall-growing subsp. *menziesii* of the Pacific "rain forest" belt, and the bluish-leaved subsp. *glauca* of the drier inland mountains. That subspecies ranges south in fragmented patches well into Mexico.

The Douglas Fir grows rapidly, but ring counts show it can reach up to 1300 years of age. Old-growth lowland forests are now nearly all felled for timber, used worldwide for building.

English Oak

Quercus robur

Height: up to 120 ft (37 m)
Type: winter-deciduous broadleaf
Occurrence: most of Europe eastward to Russia; Caucasus; Turkey; northwest Africa
Habitat: temperate deciduous forest and woodland, on moderately fertile soils
Division: Magnoliophyta: dicotyledons **Family:** Fagaceae

The oaks (*Quercus*) are a genus of over 400 species. Most are evergreens, and the greatest diversity occurs in the tropics and northern hemisphere temperate zones. English Oak is a somewhat parochial name for this well-known species. It is much more abundant in France and Germany, where large oak forests are still managed for long-term timber yield. However, it has had a special significance for England, as it was ships built of this oak that gave the country naval supremacy for most of the four centuries up to and even beyond the Napoleonic Wars. However, by the late 17th century, English Oak forests were so depleted there that most supplies were shipped from Baltic countries.

The English Oak has a relatively short bole and a broad canopy. Its short-stalked leaves are lobed, and the acorns are long-stalked. Some specimens are more than 1000 years old.

Cork Oak

Quercus suber

Height: up to 70 ft (21 m)
Type: evergreen broadleaf
Occurrence: western Mediterranean; nearby Atlantic coast of Europe and northwest Africa
Habitat: evergreen oak and pine woodland and scrubland, mostly on rocky hills
Division: Magnoliophyta: dicotyledons **Family:** Fagaceae

The Cork Oak is just one of many small-leaved evergreen oak species, most of them from southern China, Japan, Mexico and California, USA, where they dominate scrubby hills. What makes this oak special is the properties of its thick bark. This is the pure cork used to seal bottles of quality wine, as well as having a myriad of other uses. It was well known to the ancient Romans, whose word *suber* applied both to the tree and the cork. It may have derived from the Greek *syphar,* meaning "wrinkled skin," referring to the bark's wrinkled outer surface. The water-repellent compound suberin is found in the bark cells of many trees, but in the Cork Oak it is in a uniquely pure form.

Limbs of a Cork Oak show its rugged bark and small leaves. A cylinder of outer bark can be harvested every 10 years without harming the tree. Portugal is the largest cork producer.

Red Mangrove

Rhizophora mangle

Height: up to 80 ft (24 m)
Type: evergreen broadleaf
Occurrence: tropical and subtropical Americas, north to southern Florida
Habitat: intertidal zone in bays, estuaries and creeks, on sand or mud deposits
Division: Magnoliophyta: dicotyledons **Family:** Rhizophoraceae

Mangle, the Spanish name for this tree, was taken from its name in Taino, the now almost extinct language once widely spoken in the Caribbean. It has also given rise to the English "mangrove." The word refers to a variety of trees, from a number of unrelated plant families, that are adapted to survive in more saline environments than normal plants will tolerate. *Rhizophora* is a major mangrove genus in the tropics. Like some other mangroves, it is a "live bearer." That is, the seeds germinate while still inside the fruit attached to the parent tree. The elongated seedlings then drop into the water and drift like fishing floats until they come to rest and take root on a muddy or sandy shore.

The Red Mangrove is the largest American mangrove. It has dramatically arched stilt roots that balance the tree in soft sand or mud and absorb oxygen through their exposed surfaces.

Indian Rhododendron

Rhododendron arboreum

Height: up to 100 ft (30 m)
Type: evergreen broadleaf
Occurrence: Himalayas from Kashmir to southwest China and north Vietnam; southern India; Sri Lanka
Habitat: coniferous and broadleaf (evergreen oak) forest on steep mountain slopes
Division: Magnoliophyta: dicotyledons **Family:** Ericaceae

After roses, no group of shrubs is better loved in temperate gardens than the rhododendrons. Most of those planted are hybrids, but they have been bred from a much more diverse genetic base than roses. *Rhododendron* has more than 1000 species, including the azaleas, and ranges from northeastern Australia to above the Arctic Circle. There are only about 30 in the Americas. Most are in eastern Asia, and *R. arboreum* is one of the most wide-ranging Asiatic species. Its regional forms are recognized as five subspecies. It was the first Himalayan rhododendron brought to Europe (in 1826), and was soon crossed with species from America and the Caucasus to produce many cold-hardy hybrids.

The Indian Rhododendron seldom grows more than 25 ft (8 m) tall in gardens, but is much taller in native forests. Many wild stands are depleted by grazing and firewood gathering.

Weeping Willow

Salix babylonica

Height: up to 60 ft (18 m)
Type: winter-deciduous broadleaf
Occurrence: northern and central China
Habitat: riverbanks and sandbars, in disturbed woodland and scrubland on alluvial soils
Division: Magnoliophyta: dicotyledons **Family:** Salicaceae

Rather atypical of the large willow genus *Salix*, the Weeping Willow has been the subject of uncertainty regarding its origins and classification. It was not known in Europe before 1700 and was thought to have come from the Middle East. Linnaeus named it *S. babylonica* in 1753 in the mistaken belief that it was the tree beneath which the Hebrew slaves in Babylon sat and wept "when we remembered Zion." But "willow" in the Bible was a mistranslation of the name of another tree, *Populus euphratica*. It now seems that it is a weeping clone of a variable species, also known as *S. matsudana,* that was perhaps carried over the Silk Road from its native northern China. Another popular mutant form is the Corkscrew Willow (cultivar 'Tortuosa').

The Weeping Willow grows rapidly, reaching its mature size in 20 years. It prefers climates with warm, dry summers, and grows best in permanently wet sites such as streambanks.

Japanese Umbrella Pine

Sciadopitys verticillata

Height: up to 120 ft (37 m)
Type: evergreen conifer
Occurrence: southern Honshu, Shikoku and Kyushu, Japan
Habitat: mixed coniferous and broadleaf forest in high-rainfall mountain areas at 2000–5000 ft (610–1520 m)
Division: Pinophyta **Family:** Sciadopityaceae

Ever since this remarkable Japanese conifer became known to science, there has been debate about where it belongs in classification schemes, because its morphology is quite unique. Now it is generally agreed that it is a truly ancient relic. Fossils as far back as the Triassic period (about 160 million years ago) show little difference from living trees but are more widespread across the northern hemisphere. Molecular evidence also supports its uniqueness, with no close relatives among the conifers. Its most striking feature is its foliage—widely spaced whorls of blunt-tipped needles up to 4 in (10 cm) long. This structure has mystified botanists, who are uncertain whether they are pairs of true leaves fused back-to-back or modified photosynthetic branches.

The name *Sciadopitys* comes from the Greek *skiadeion,* for "umbrella," and *pitys,* for "pine," alluding to the foliage's needle arrangement. There are some famous ancient trees in Japan.

Coast Redwood

Sequoia sempervirens

Height: up to 378 ft (115 m)
Type: evergreen conifer
Occurrence: California north from Monterey County, and far southeast Oregon, USA
Habitat: moist coniferous forest on coastal plains and foothills less than 25 miles (40 km) from ocean, in maritime "fog belt"
Division: Pinophyta **Family:** Cupressaceae

There is little doubt that this species currently includes the world's tallest trees. Even now, a few enthusiasts are combing the choicest redwood groves in northern California for new record trees, measuring them accurately by the "tape drop" method, which requires climbing the tree to the very top. The latest record is 378.1 ft (115.25 m), for a tree measured in 2006 in Redwood National Park. *Sequoia,* a genus of this single species, became known to science in 1796 through Archibald Menzies' collections. It was initially called *Taxodium sempervirens,* and was not named as a distinct genus until 1847. The name was apparently inspired by the Cherokee leader and scholar Sequoyah.

A Coast Redwood stand shows the perfectly straight trunks and thick, fibrous bark, which protects the living tissue from fire. Most old-growth stands are now protected from logging.

Giant Sequoia

Sequoiadendron giganteum

Height: up to 300 ft (91 m)
Type: evergreen conifer
Occurrence: western slopes of Sierra Nevada, central California, USA
Habitat: coniferous forest in moist sites in valleys at 3000–7000 ft (910–2130 m)
Division: Pinophyta **Family:** Cupressaceae

Californians use the common names Redwood and Sequoia in a consistent way, but it may confuse outsiders. Their Giant Sequoia is in fact *Sequoiadendron,* while the Redwood or Coast Redwood is *Sequoia.* The Giant Sequoia is the largest tree in terms of timber volume in the world because of its great height and a trunk diameter of more than 20 ft (6 m). The current champion is the General Sherman tree at 275 ft (83.82 m) tall, 27 ft (8.23 m) in diameter, and 130 ft (39.6 m) to its lowest major branch. Until about 60 years ago, a Giant Sequoia was thought to be the world's oldest tree, with 3200 annual growth rings. But the Bristlecone Pine (*Pinus aristata*) is now known to grow older.

This is the view up into the crown of the General Sherman tree. The fire-resistant orange brown bark can be up to 2 ft (60 cm) thick. These grand trees attract many tourists.

Sal

Shorea robusta

Height: up to 120 ft (37 m)
Type: dry-season-deciduous broadleaf
Occurrence: northern and central India; Nepal; Bangladesh; Burma
Habitat: tropical deciduous forest on hills up to 5000 ft (1520 m) in monsoonal regions
Division: Magnoliophyta: dicotyledons **Family:** Malvaceae

The Sal is India's most important native timber tree. Also, it is a representative of the dipterocarps, a large group of tropical Asian trees formerly treated as a family in its own right but now understood as a branch that arose within the broader mallow family (Malvaceae). Dipterocarps are characterized by fruit that have two or more persistent, enlarged sepals, like propeller blades. These allow the fruit to spin slowly down from the tree and so be carried on a breeze for some distance, aiding seed dispersal. In much of the wet Asian tropics, dipterocarps are the dominant rain forest tree and foremost among exploited rain forest timbers. There are almost 700 species, 360 of them in *Shorea*.

Open-grown Sal has a crooked, bumpy trunk, but forest-grown Sal has a long, straight trunk. The strong timber exudes a resin, *dammar*. Sal-butter, from the seeds, is used in cooking.

Mahogany

Swietenia mahogani

Height: up to 100 ft (30 m)
Type: dry-season-deciduous broadleaf
Occurrence: tropical Americas, especially Caribbean Islands; far south Florida
Habitat: tropical rain forest, vine thickets and scrubland, often on swamp margins or lowland hills
Division: Magnoliophyta: dicotyledons **Family:** Meliaceae

Mahogany was prized from the 16th century onward for its timber, which combines strength, hardness, durability, closeness of grain and ease of working to a degree matched by few others. Spanish colonists exploited it ruthlessly and large quantities were exported—mainly for shipbuilding at first, but by the mid-18th century it was the timber of choice for fine furniture. This species is often called West Indian Mahogany, to distinguish it from the taller Honduras Mahogany (*S. macrophylla*) of Central and South America. "Mahogany" is thought to come from *m'oganwo*, the Yoruba name for the related African *Khaya senegalensis*, which slaves transported from Nigeria gave to this Caribbean tree.

After centuries of logging, most remaining wild Mahogany are small, crooked trees, or even shrubs if on exposed sites. Widely planted in parks, it develops a broad, shady canopy.

Clove

Syzygium aromaticum

Height: up to 60 ft (18 m)
Type: evergreen broadleaf
Occurrence: somewhere in the Moluccas, eastern Indonesia; uncertain due to long cultivation
Habitat: lowland tropical rain forest
Division: Magnoliophyta: dicotyledons **Family:** Myrtaceae

Cloves were one of the prized spices traded since ancient times through a long chain of seafarers and merchants between the "Spice Islands" of the East Indies and Europe. They are the dried flower buds of this species of *Syzygium*, which is one of largest tree genera of the world, with about 1000 species. Most *Syzygium* species have quite low concentrations of essential oils, even by comparison with other members of the myrtle family. The Clove is an exception. Its powerful oil, eugenol, concentrated in the drying buds, has both antiseptic and anesthetic properties as well as a penetrating flavor. It is still one of the best painkillers for toothache. It is also the additive that gives Indonesian *kretek* cigarettes their distinctive smell.

Clove plantations were forbidden outside a few islands in the Moluccas by the 17th-century Dutch, to preserve their monopoly. They were later established in parts of the wet tropics.

Tamarind

Tamarindus indica

Height: up to 60 ft (18 m)
Type: evergreen broadleaf
Occurrence: tropical Africa or western Asia; uncertain due to long cultivation
Habitat: monsoonal woodland, savanna and scrubland, mostly on plains
Division: Magnoliophyta: dicotyledons
Family: Fabaceae (Leguminosae)

This is another tropical tree so widely planted for so long that its wild origins cannot readily be traced. A small, leguminous tree with ferny, pinnate leaves, it produces abundant plump pods with brittle husks that enclose a row of large, hard seeds embedded in sticky, sour-sweet brown pulp. The pulp is important in many cuisines, especially in Asia, and is used for beverages and bottled sauces too. The fallen pods provide food for browsing animals, both wild and domesticated, who spread the seeds in their dung. Humans have also carried seeds long distances. "Tamarind" derives from the Arabic for "Indian date."

The Tamarind's pods ripen in the tropical dry season. With adequate soil moisture and fertility, it makes a fine tree with a dense, shady canopy. In drier sites it is small and crooked.

Athel Tree

Tamarix aphylla

Height: up to 50 ft (15 m)
Type: evergreen broadleaf
Occurrence: northeastern Africa; western Asia
Habitat: banks and channels of nonperennial streams in arid regions, mostly in sandy or silty soils
Division: Magnoliophyta: dicotyledons **Family:** Tamaricaceae

At first glance, an Athel Tree looks like a conifer of the cypress kind, with minute scale leaves pressed against delicate, needle-like twigs. But it is a true flowering plant, producing panicles of small, pink flowers through summer and masses of small, silk-down seeds in autumn. The Athel (from its Arabic name) Tree has long been valued in its native regions for its ability to thrive in desert climates, giving dense shade, shelter from wind, and firewood. It was widely planted in other continents for the same reasons, but has become a significant weed in the southwestern USA and central Australia, spreading along watercourses and displacing native vegetation. This is the tallest of the 55 species of *Tamarix*. The others are shrubs or small trees.

The tiny grayish leaves of the Athel Tree are arranged spirally on very slender branchlets. The five-petaled flowers are very small too. It branches low, with thick limbs and furrowed bark.

Bald Cypress

Taxodium distichum

Height: up to 150 ft (46 m)
Type: deciduous conifer
Occurrence: southeastern USA from Florida to Delaware, Illinois and eastern Texas
Habitat: freshwater swamps and low-lying riverbanks, often in quite deep water
Division: Pinophyta **Family:** Cupressaceae

The Bald Cypress is as renowned in popular lore as it is among botanists and tree lovers. Who has not been impressed by film sequences shot in the gloomy "great cypress swamps" of Florida or Louisiana? Such swamps are dominated by this tree species, one of only two in the *Taxodium* genus. It is unusual, but not unique, among conifers in being deciduous. Very small leaves are attached in two comblike rows to delicate branchlets. In fall, the branchlets and leaves turn shades of gold to deep russet, then are shed as a unit. The Bald Cypress has a remarkable feature, its "knees." These woody, upward-growing projections from the root system aid the exchange of gases with the air.

A Bald Cypress in fall color stands in a typical cypress swamp, draped with Spanish Moss (a pineapple relative, not a moss). This tree can grow just as well in normal garden soil.

Montezuma Cypress

Taxodium mucronatum

Height: up to 120 ft (37 m)
Type: deciduous conifer
Occurrence: Guatemala; Mexico; Rio Grande valley, southern Texas, USA
Habitat: freshwater swamps, riverbanks and canyons, and beside lakes, springs and soaks, on plateaus
Division: Pinophyta **Family:** Cupressaceae

This is Mexico's national tree, famous for the huge trunk girth of specimens in the central highlands. Trees in Mexico City's Chapultepec Park are among the tallest and are said to have been planted in the time of the last Aztec rulers. Montezuma Cypress is the English name. The Mexican name is *ahuehuete* (a-weh-wah-de), from the Nahuatl language. Botanists have long debated the status of this tree in relation to its northern relative, the Bald Cypress (*Taxodium distichum*). The leaves, twigs and cones of both are virtually identical, and it has been argued that they are a single species. But the growth forms of the two trees are quite distinct. Also, the Mexican tree is often almost evergreen and lacks "knees."

El Arból del Tule in Oaxaca State, Mexico, is 115 ft (35 m) tall, with trunk almost 40 ft (12 m) in diameter. Its bulk rivals that of the largest Giant Sequoias (*Sequoiadendron giganteum*).

Yew

Taxus baccata

Height: up to 100 ft (30 m)
Type: evergreen conifer
Occurrence: Europe; northwest Africa; temperate western Asia east to northern Iran
Habitat: mixed woodland and scrubland, usually on rocky hillsides, ravines or outcrops
Division: Pinophyta **Family:** Taxaceae

The cultural significance of the Yew is so great and varied that its botanical aspects have been overshadowed. It is one of about seven species of the yew genus (*Taxus*), although their botanical and chemical characteristics are so similar that they could arguably be treated as a single species. They range in a fragmented pattern through the northern hemisphere, and just into the southern hemisphere in the mountains of Sulawesi. The Yew's status as a conifer is not obvious, as each cone is reduced to a single seed surrounded by a highly modified scale that is cup-shaped, red and juicy. It is eaten by birds, who thus distribute the seed. All other parts of yews are poisonous but contain the anticancer compound taxol.

The Yew is often associated with churches and graveyards. Trees with trunks up to 10 ft (3 m) in diameter are known, with ages calculated at more than 2000 years old.

Teak

Tectona grandis

Height: up to 130 ft (40 m)
Type: dry-season-deciduous broadleaf
Occurrence: mainland tropical Asia from central India to Vietnam
Habitat: tropical deciduous forest and rain forest margins, in fertile soils on lower hill slopes up to 4000 ft (1220 m)
Division: Magnoliophyta: dicotyledons **Family:** Lamiaceae

Teak timber is renowned for its highly useful properties. It is dense, strong, easily worked, resists splitting and splintering, is highly durable when exposed to weather, and contains an oily, waxy material that inhibits rusting of iron bolts and nails. These qualities were well known to Asian carpenters and shipbuilders long before Europeans used Teak. Although not native in Indonesia, it was established there about 500 years ago for boat-building. In the mid-18th century the British Navy began to use it, initially from shipyards in Bombay. By the end of the Napoleonic Wars, it was the favored timber for warships' decks, rails, gun carriages and outer planking. Many injuries resulted from flying splinters of oak during sea battles; fewer came from Teak.

Nowadays, Teak is used mainly for outdoor furniture. Most supplies have previously come from Burma, but large plantations in other Asian countries are now yielding good timber.

Cacao

Theobroma cacao

Height: up to 25 ft (8 m)
Type: evergreen broadleaf
Occurrence: Central America; northern South America on lower slopes of Andean mountains
Habitat: understory of tropical rain forest, mainly in disturbed margins
Division: Magnoliophyta: dicotyledons **Family:** Malvaceae

Perhaps the greatest gift from pre-European Mesoamerican culture to the world, the product of this tree is now one of the foremost commodities in international trade. The Cacao was shown in the wall carvings of the Maya, Tolmec and Aztec. It and its uses were reported in detail by early Spanish colonists, who also rendered the Nahuatl name for the beverage, *xocolatl,* into Spanish, *chocolate*. This was soon adopted by other European languages. The name Cacao, also from Nahuatl, is reserved for the tree and its raw product, the "beans." However, it too has come into English as "cocoa." Linnaeus coined the botanical name *Theobroma,* meaning "food of gods."

The Cacao is a large-leaved, shade-loving tree. It bears pods up to 1 ft (30 cm) long from the larger branches. The large, fatty seeds are fermented, dried and roasted before processing.

American Elm

Ulmus americana

Height: up to 120 ft (37 m)
Type: winter-deciduous broadleaf
Occurrence: eastern North America, from Saskatchewan and Montana to central Texas
Habitat: broadleaf deciduous forest on deeper soils in valley bottoms and along riverbanks and lake margins
Division: Magnoliophyta: dicotyledons **Family:** Ulmaceae

The American Elm is one of about 30 species of the elm genus (*Ulmus*), which are distributed through temperate regions of the northern hemisphere. Nearly all are deciduous, with simple, toothed leaves that tend to be asymmetric at the base. The small, inconspicuous flowers are soon followed by greenish, winged fruit that are blown about by wind, thus distributing the seed. Elms have been devastated by Dutch elm disease which is carried by a fungus transmitted by elm bark beetles. It killed countless elms in Europe in the 1930s and the 1960s, and also spread to America. Before that, the American Elm was regarded as one of the finest native trees for street and park planting, as well as for timber.

This mature American Elm displays its characteristic high, umbrella-shaped crown. Such specimens are now rare because of the spread of Dutch elm disease.

Wollemi Pine

Wollemia nobilis

Height: up to 130 ft (40 m)
Type: evergreen conifer
Occurrence: one limited area in Wollemi National Park, northwest of Sydney, eastern Australia
Habitat: temperate rain forest on organic alluvium in bottom of narrow sandstone canyon
Division: Pinophyta **Family:** Araucariaceae

The discovery of this amazing new conifer in 1994 excited keen gardeners, naturalists and conservationists around the world. David Noble, an explorer of canyons in the sandstone wilderness of the Blue Mountains region west of Sydney, brought back the twig of a strange, unknown tree from a remote and inaccessible canyon. Botanists made further collections and soon realized that it represented a new, third genus of the ancient Araucariaceae family. It also proved to be a good match for fossils from the Cretaceous period, more than 100 million years old. Quite different from any other conifer, it has now been propagated in large numbers and is surprisingly hardy.

A young Wollemi Pine displays some unique features. Leaves are spirally arranged on twigs but twisted at the bases to lie in two rows. On later twigs, this changes to four rows.

Grass Tree

Xanthorrhoea glauca

Height: up to 26 ft (8 m)
Type: evergreen swordleaf
Occurrence: southeastern Australia
Habitat: woodland or shrubland on steep ridges and sides of gorges, in shallow, rocky soil
Division: Magnoliophyta: monocotyledons
Family: Xanthorrhoeaceae

Xanthorrhoea is an endemic Australian genus of 28 species, only a few of which grow substantial aboveground stems. This species, one of the tallest, occurs in large numbers on rocky ranges in southeast Australia. Grass Trees have a unique structure, with a true stem of fibrous conducting tissue supported by a sheath of tightly packed old leaf bases glued by a reddish crystalline resin. Hundreds of very narrow, hard-textured leaves radiate from the apex of each branch. Tall, rodlike flower spikes grow above the foliage, then numerous tiny, white flowers emerge from densely packed, brown bracts. Nectar-rich, they attract many insects. The plants are highly fire-resistant, and are among the first to resprout after wildfire.

Grass Trees grow on a cliff edge in rugged sandstone land in Australia. The flower spikes consume much of the plant's energy store and may not reappear for several years.

Joshua Tree

Yucca brevifolia

Height: up to 50 ft (15 m)
Type: evergreen swordleaf
Occurrence: Mojave Desert, southeastern California and adjacent corners of Arizona, Nevada and Utah, USA
Habitat: arid woodland and desert scrubland, on plateaus at 1500–6500 ft (460–1980 m)
Division: Magnoliophyta: monocotyledons **Family:** Agavaceae

After the Saguaro (*Carnegiea gigantea*), the Joshua Tree is the plant that many people associate with the USA's southwestern deserts. The two occur in slightly different areas—the Joshua Tree in the Mojave Desert uplands, the Saguaro farther southeast in the Sonoran Desert. The Joshua Tree is the tallest of about 30 species of *Yucca,* a genus confined to North America and Mexico, where most other large tree yuccas are found. All yuccas have tough, spine-tipped leaves arranged in close spirals on the branches. At intervals they produce showy panicles of white, bell-shaped flowers on the branch tips. The epithet *brevifolia* refers to the unusually short leaves, mostly less than 1 ft (30 cm) long.

Joshua Tree National Park, at the foot of the San Bernardino Mountains not far from Los Angeles, is where some of the largest, and oldest, Joshua Tree specimens can be seen.

Communities of life

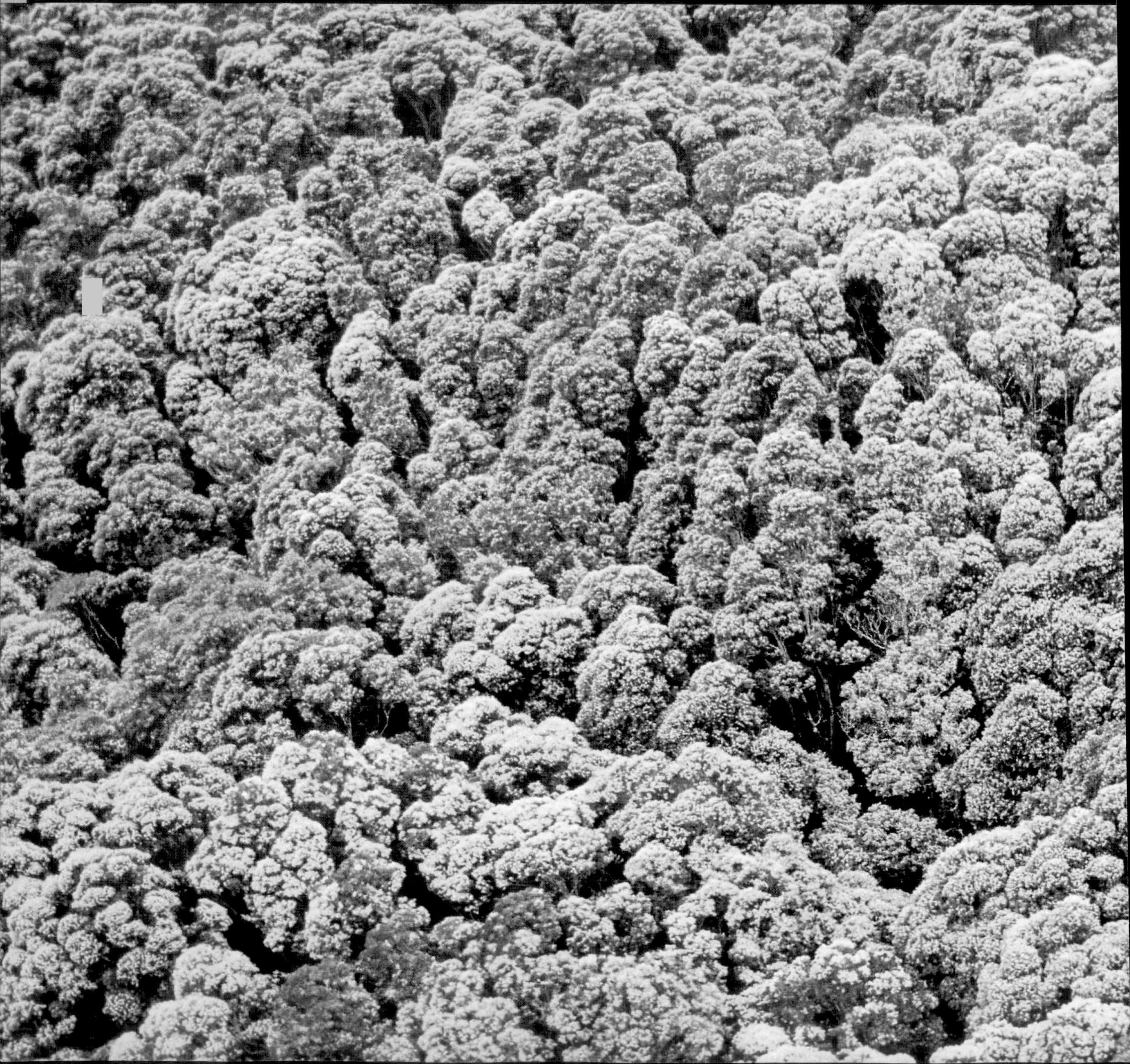

Previous page The lush green temperate rain forest canopy in Westland National Park on New Zealand's South Island.

Communities of life

Trees dominate our natural world and their presence, in the form of forest type, largely defines our climate zones. Biologists have identified Earth's various "biomes," which are communities of plants, animals and soil organisms, all of which rely on the trees for shelter, moisture and food.

Great forests of the world 180
Boreal forest 182
Temperate coniferous forest 186
Temperate broadleaf deciduous forest 190
Temperate southern hemisphere broadleaf forest 194
Temperate woodlands 198
Tropical rain forest 202
Tropical dry deciduous forest 206
Savanna 210
Mangroves 214
The urban landscape 218

Great forests of the world

The natural vegetation covering the major part of the world's land area (excluding Antarctica) is dominated by trees. It is the presence of trees which defines, in descending order of tree density: forest, woodland and savanna. Admittedly there are large regions of the world where humans or their domesticated animals have removed most of the trees over the last 1000 or more years; but in many such regions there is evidence that trees will rapidly recolonize when grazing or agriculture is discontinued—such as resulting from the "flight to the cities" that has occurred in many impoverished, or once-impoverished, societies. Tree cover varies hugely from one part of the world to another. It is largely climate that determines its height, structure, density and species composition. Foremost among climatic variables are temperature, including both summer and winter mean and extreme temperatures; rainfall, both as to annual total and its distribution through the seasons; humidity; and wind.

CLASSIFYING TREE VEGETATION

Trees are the largest land organisms: where present in large numbers they dominate and modify the environment, providing shelter, moisture and food for a myriad of other organisms. As the forest or woodland type changes, so too does the whole assemblage of dependent organisms. Plant ecologists classify vegetation according to several criteria: the most important being structure, which encompasses features such as tree density, canopy height, presence of lower layers such as understory trees, shrubs and grasses, and presence of growth forms such as lianes, epiphytes or parasitic plants; and species composition, meaning numbers and identification of species of trees as well as all other plants. At the highest level of classification, biologists recognize biomes—broad-brush groupings of the world's biota (totality of organisms)—though, as noted above, these are essentially determined by type of tree cover, at least for regions where trees are present.

MAJOR FOREST BIOMES OF THE WORLD

- Boreal forest
- Temperate coniferous forest
- Temperate broadleaf deciduous forest
- Temperate southern hemisphere broadleaf forest
- Temperate woodlands
- Tropical rain forest
- Tropical dry deciduous forest
- Savanna
- Mangroves

↓ **Forest and woodland in Slovenia,** eastern Europe. It is likely that this scene includes some trees that have recolonized neglected farmland, a result of increasing urbanization.

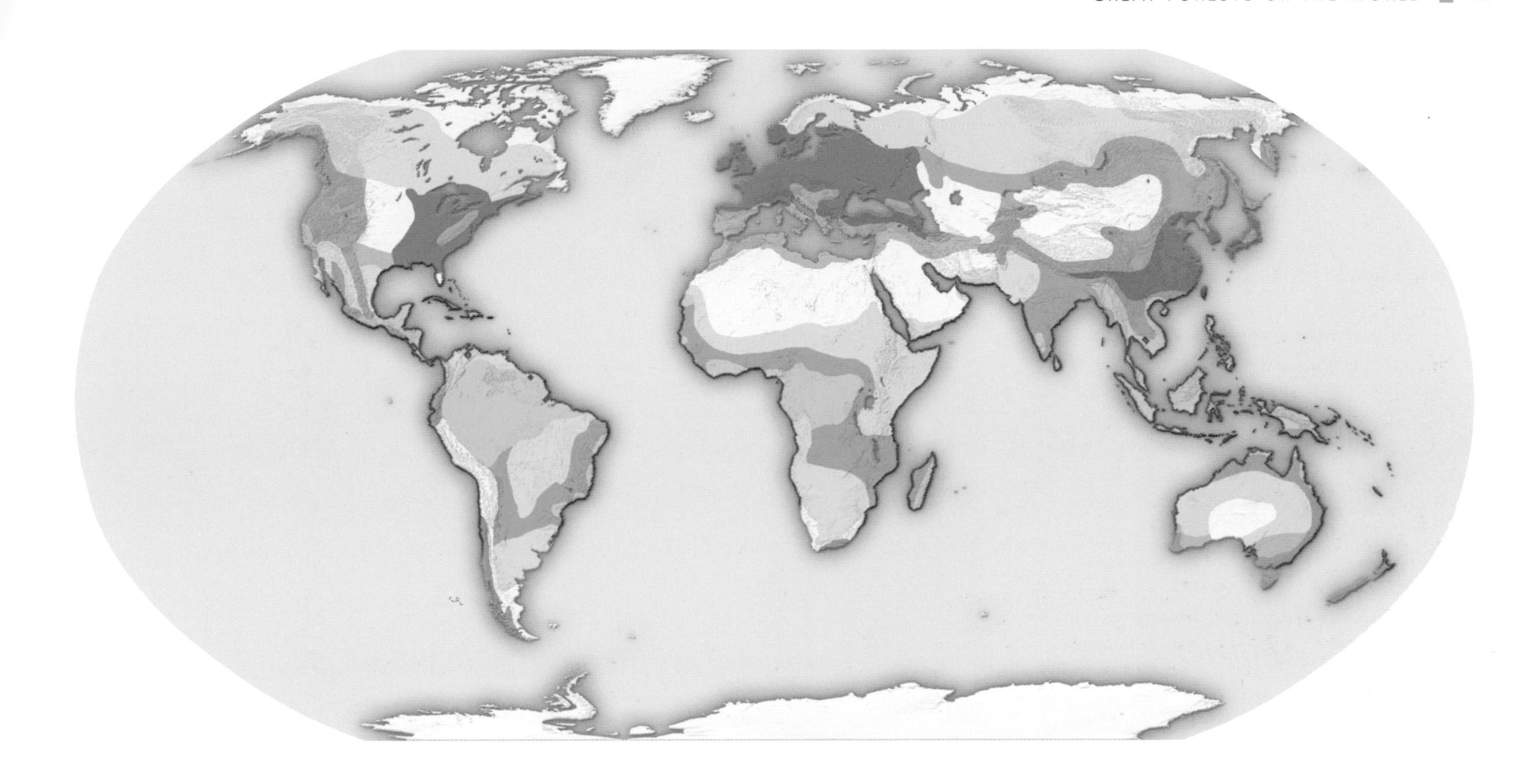

Boreal forest

Boreal means "of the north" and it is around the cold northern regions of the two great continents of Eurasia and North America that this forest type extends. Alternatively known by its Russian name *taiga*, it is so vast that it is estimated to contain over one third of all the world's trees—not much reduced in number in recent centuries since the human population is so sparse here and has had less impact than in most warmer regions. Lumber harvesting has taken place over large tracts of these forests, but young trees replace those felled; and a large proportion still remains as wilderness. A major threat now looming is global warming. Beyond the northern limit of boreal forest lies the treeless arctic tundra. The latitude of this limit varies, being at most 2–3 degrees above the Arctic Circle. Boreal forest stops short of most parts of the Atlantic and Pacific coasts, where ocean currents keep winter temperatures somewhat warmer—an exception is east-coastal Canada and Newfoundland, where the cold Labrador Current pushes away the warmer Gulf Stream. Boreal forests dip farthest to the south near the colder centers of the continents, and in bleak mountain areas, such as the northern Appalachian chain. As you might expect, the common feature of boreal forest is extreme cold and a long winter season. The soil freezes, thawing for only a few months in the farthest northern forests but for 7–8 months in the more southern parts. Winter air temperatures can fall as low as –80°F (–62°C) in winter.

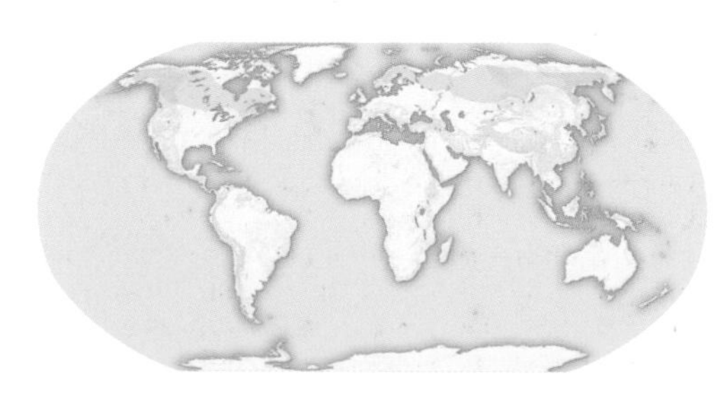

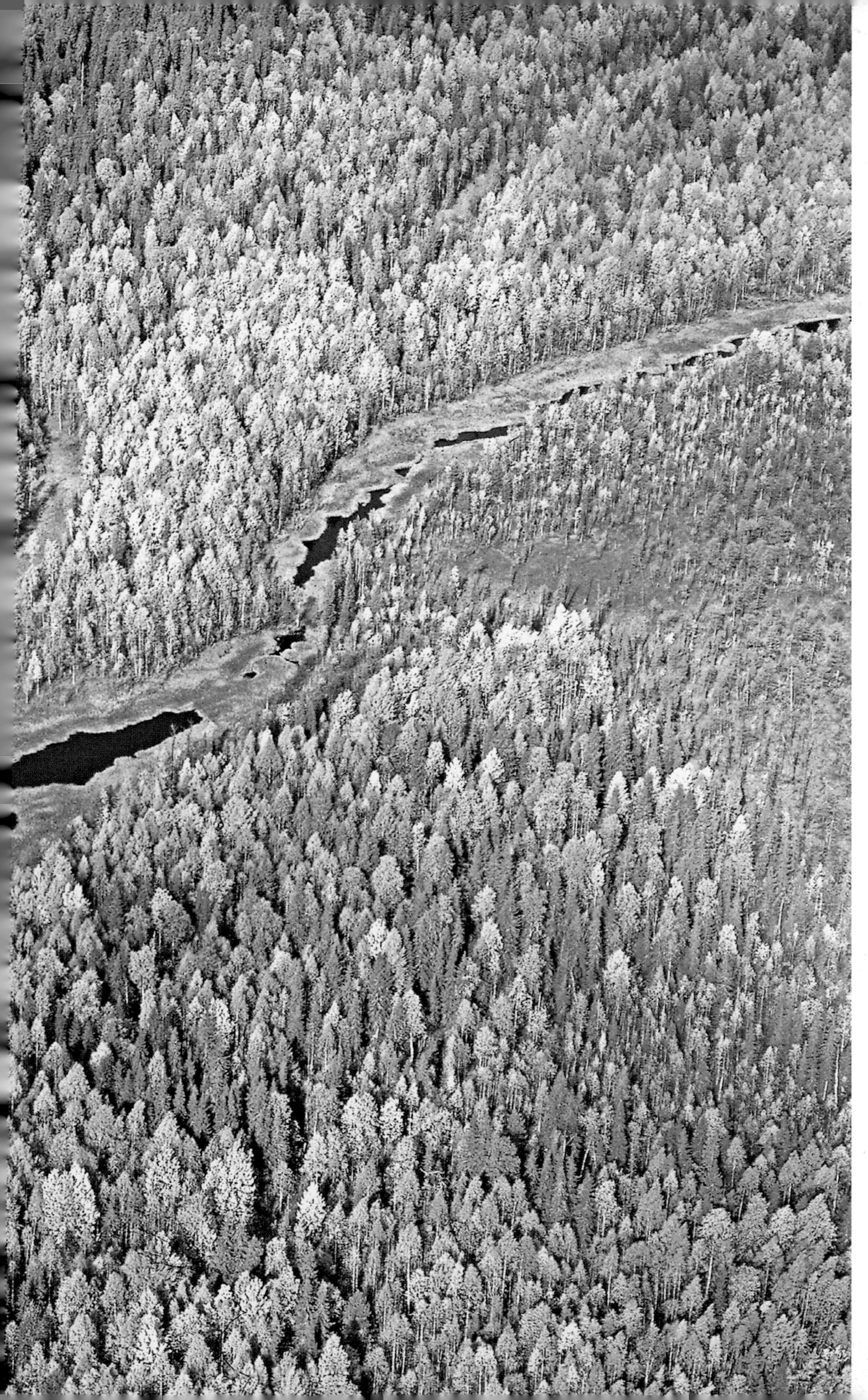

↑ **Young seed cones and new season's needles** appear together on branch tips of the deciduous Tamarack (*Larix laricina*). Tamarack is the most wide-ranging of the American larches. It occurs through most of the boreal forest zone but only becomes dominant on some boggy sites.

← **Early autumn in the taiga,** at the Pechoro-Ilychsky Nature Reserve in Russia's northern Urals. Some of the most pristine boreal forest in Eurasia is preserved here. Evergreen conifers include spruce (*Picea abies*), fir (*Abies sibirica*) and pine (*Pinus sibirica* and *P. sylvestris*).

CONDITIONS IN THE BOREAL FOREST

In the northern parts of the boreal forest zone there is an extensive occurrence of permafrost, that is, a permanently frozen soil zone at varying depths below the soil surface. Only the "active zone" of soil above the permafrost, which thaws in summer, can support plant life, and this must be of a certain depth before it can support trees. The summers are brief but can be quite hot and humid. The combined annual rainfall and snowfall is generally quite low—as little as 6 inches (150 mm) in central Alaska—but evaporation is also low, so the ground is often waterlogged over large areas in summer, and surface water is plentiful. Because of the temperature and soil limitations, boreal forests have low access to plant nutrients and tree growth is slow. There is also a smaller number of plant species than in forests in warmer climates, and most species have wide geographic ranges.

Boreal forest continued

The dominant trees of the boreal forest are conifers, belonging to just four genera in the family Pinaceae: *Abies* (fir), *Larix* (larch), *Picea* (spruce) and *Pinus* (pine). No species of these genera occurs in both Eurasia and North America, but some are closely allied to their counterparts in the other continent. Conifers, more than any other trees, have evolved adaptations to survive very low temperatures; their leaves are needle-like with a small surface area, and the concentration of sugars and starches in their sap increases as winter approaches; together with specialized proteins these form a natural "antifreeze"; oils and resins, present in numerous ducts through leaves, bark and wood, are also protective; and the dormant winter buds at the growing tip of each branch are each coated in a blob of resin. Virtually the only broadleaf trees of boreal forest are species of *Betula* (birch), *Alnus* (alder), *Populus* (poplar), *Salix* (willow) and *Sorbus* (rowan). All are deciduous, producing leaves in the brief summer to rapidly photosynthesize and so build up food reserves for the winter and spring. Low shrubs, both deciduous and evergreen, may be quite diverse; they include many species of dwarf willow (*Salix*) as well as many members of the heath family, Ericaceae.

↓ **The American black bear** (*Ursus americanus*) shares some parts of its boreal forest range with the grizzly bear (*U. arctos*), which has been known to include black bears in its diet. The black bear feeds mainly on herbage, fruit and nuts but is an occasional carnivore. Its coat color varies between geographical regions.

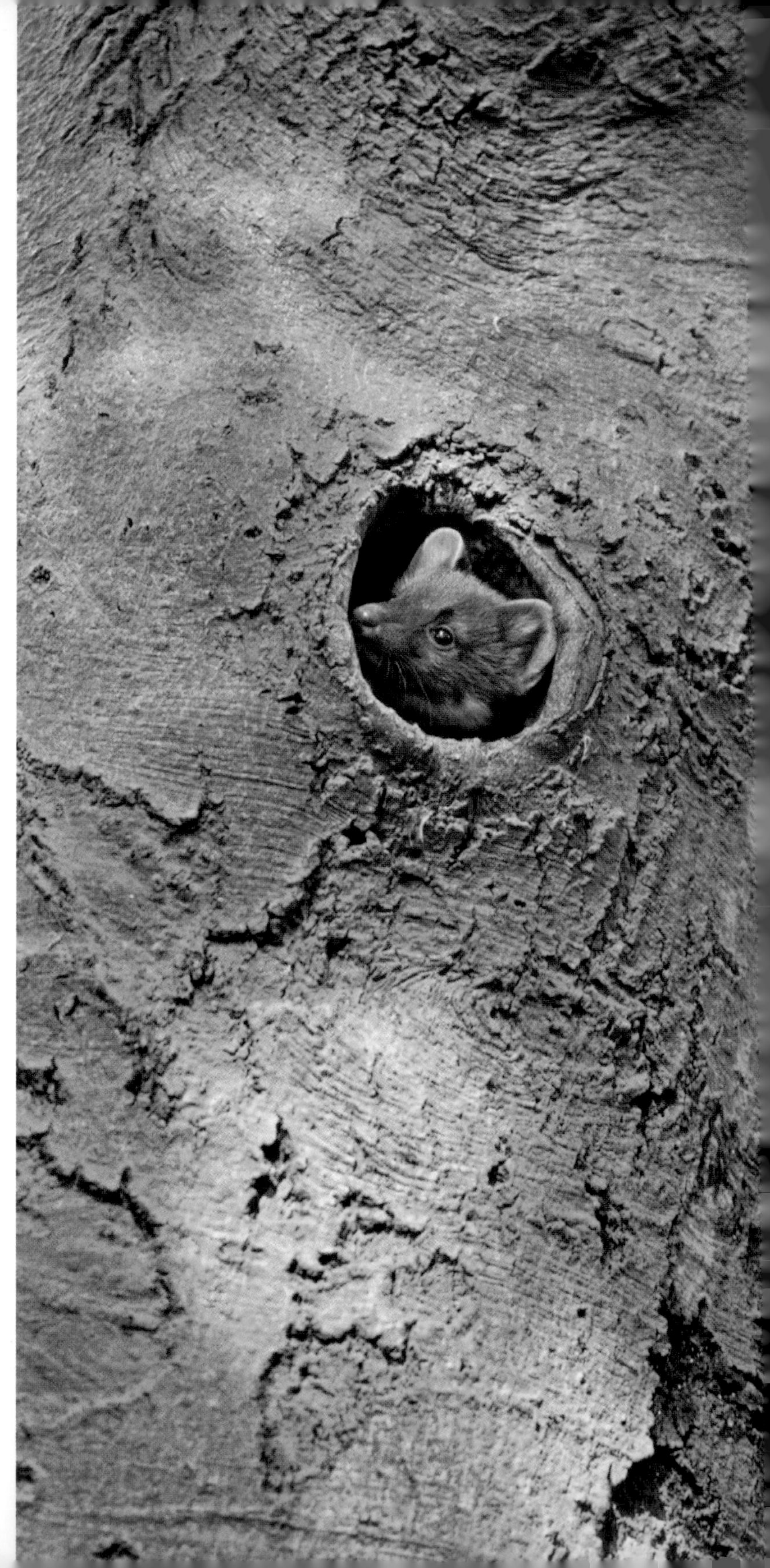

↑ **The North American porcupine** (*Erethizon dorsatum*) is an excellent tree climber. Being mainly active at night, it spends most of its days in a tree, sheltering in a suitable den.

↗ **One of the smaller owls,** only 8–12 inches (20–30 cm) long, the boreal owl (*Aegolius funereus*) ranges across both North America and Eurasia in the denser forests of the more southern parts of the boreal forests. It preys mostly on small rodents.

← **A young pine marten** peeps from its nest hollow in a moss-covered pine trunk. A carnivore of the weasel family, its preferred prey is red squirrels but it can be omnivorous.

Woodpeckers tap holes in conifer trunks with their powerful beaks, and then extend their long tongues into the crevice in order to reach the larvae of beetles and other insects lurking underneath the bark.

LIFE IN THE BOREAL FOREST

Boreal forests support a diverse fauna with complex food chains. Insects, feeding both on vegetation and other small fauna, breed prolifically in summer and are eaten in large numbers by birds, small mammals, frogs and other insects. Other mammals and birds are grazers and browsers, feeding on summer grasses and herbs, tree and shrub foliage, or in winter on lichen ("reindeer moss") scraped from beneath the snow or on the sugar-rich inner bark and twigs of trees; some birds and smaller mammals such as squirrels specialize in eating the seeds of conifers. The largest herbivores include the caribou (known as reindeer in Europe), the moose (or elk), and the arctic bison, which survives in small numbers in Alaska and Canada but is long-extinct in Siberia. Predatory mammals and birds come in all sizes, from small stoats and owls to "top predators" such as wolves and the massive grizzly bear.

Temperate coniferous forest

Within this biome are some of the world's tallest forests, namely those along the northwest coast of North America containing the Redwoods, Douglas-fir and Sitka Spruce. And not far away, in conifer forests in the California Sierras, grow the world's oldest trees, Bristlecone Pines (*Pinus aristata*), and the largest trees in terms of timber volume, the Giant Sequoia (*Sequoiadendron giganteum*). But these are the extremes of a more widespread forest type. Other major areas of temperate coniferous forest in the northern hemisphere are the mountains of western China, northeastern China and the adjacent regions of far eastern Russia, Japan, the mountains of Central Asia, the Himalayas and the Caucasus, and Mexico's Sierra Madre ranges. Central Europe, the Balkans and Turkey all have coniferous forests on their mountains, and northwest Africa has remnants of them on the Atlas ranges. Most of these are dominated by members of the pine family (Pinaceae), namely pine, spruce, fir, larch and cedar, though in west-coastal North America, trees of the cypress family (Cupressaceae) may dominate some forests. In other parts of the world there are smaller fragments of coniferous forests; even in the southern hemisphere there are the *Araucaria* and *Fitzroya* forests of Chile, the Kauri and podocarp forests of New Zealand and even some small patches in Tasmania, Australia, of ancient conifer genera.

→ **Coniferous forest clothes these hills** in northern Europe. The dominant tree is the European Spruce (*Picea abies*), which ranges from the mountains of Greece to the boreal forests of Scandinavia and Russia.

↓ **Blue Spruce** (*Picea pungens*) is a major species of the drier types of coniferous forest on the Rocky Mountain plateaus, from Idaho south to Arizona and New Mexico. It is popular as an ornamental tree.

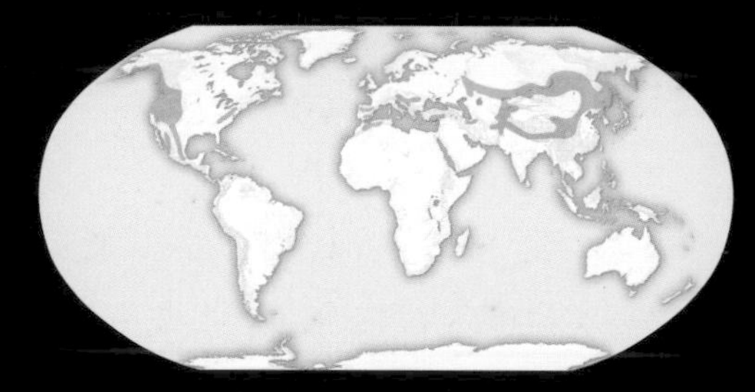

DISTRIBUTION OF CONIFEROUS FOREST
Coniferous forest occurs mainly on higher mountain areas where the summers are relatively cool and spring snow-melt provides ample soil moisture. Yet in some regions, such as west-coastal North America, some of the greatest conifer forests are found on coastal lowlands. It is likely that there are historical factors which have determined their present-day distribution, especially the survival, during the geologically recent ice ages, of many conifer genera and species in "refugia"—pockets of warmer but humid climate that escaped being obliterated by the great ice sheets. In Eurasia, such refugia were scattered around the Mediterranean, the lower Himalaya and southwest China, while in North America they were on the west coast and in the southeast. After climates warmed again and the ice sheets retreated, coniferous forests crept back up the mountain slopes over some thousands of years.

Temperate coniferous forest continued

Temperate coniferous forest is in some ways an unusual forest type with respect to the life-forms that are associated with it. Many of the plants are quite specialized in order to deal with the forest's peculiar environment. Common features of this environment, or at least of dense, well developed coniferous forest, are the low levels of light in the understory, and a close mat of fallen needles and twigs on the forest floor, that decay quite slowly and have an acidifying effect on the soil. Thus there are generally few or no broadleaf small trees and shrubs, and the sparse groundlayer plants are highly shade-tolerant. Life-forms that may be unusually well represented are epiphytic lichens, mosses and ferns, fungi, and parasitic and saprophytic flowering plants. The latter include mistletoes on upper branches as well as small ground-dwellers that lack chlorophyll and therefore have little need of light, for example the ground-cones (*Boschniakia*), pinesaps (*Monotropa*) and snowplant (*Sarcodes*). Of course where temperate coniferous forest adjoins clearings or suffers disturbance, such as from fire or landslip, a richer flora of light-loving smaller plants rapidly develops, at least temporarily. The fauna of dense coniferous forest is also quite specialized, the most abundant animals being those that live in the treetops.

→ **These monarch butterflies** overwinter in huge numbers on the trunks of the Oyamel Fir (*Abies religiosa*) on a few mist-shrouded volcanic mountains of central Mexico west of Mexico City.

↘ **Grosbeaks are large finches** with powerful beaks, and this species is a conifer specialist. Its scientific name is *Pinicola enucleator*, which means literally "kernel-extracting pine-dweller."

↓ **The fly agaric** (*Amanita muscaria*), is one of the most distinctive of all toadstools. Native to the moist pine forests of the northern hemisphere, it has become naturalized among planted pines in the southern hemisphere. Toxic, though rarely fatal, in Siberia it was used by shamans to induce religious hallucinations.

TEMPERATE CONIFEROUS FOREST LIFE
The main fauna in the temperate coniferous forest are the birds, many of these having the ability to extract the edible seeds of conifers, although others feed largely on insects, another group that is diverse in the treetops. Some small tree-climbing mammals also depend on conifer seeds, notably squirrels. Then there are many predators, which can either climb trees, such as martens and some snakes, or those that fly, such as hawks and owls. At ground level there are sparser populations of herbivores, some having fungi as a major part of their diet, others the fallen conifer seeds or "mast."

↓ **The red sqirrel,** found over much of North America, is always associated with coniferous forest. Its main food source is spruce seeds.

Using its drill-like ovipositor, the parasitic ichneumon wasp bores through the bark of a pine to lay its eggs on a sawfly grub deep inside the wood.

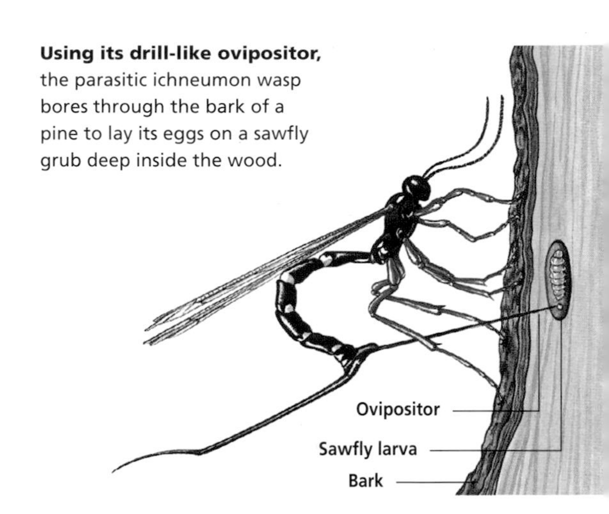

Temperate broadleaf deciduous forest

A large proportion of the people of eastern USA live in cities that were once part of this forest, as do countless millions more in China, Japan and Europe. Climatically, these forests occur in regions with hot, humid summers, cold winters and summer rainfall dominance, though with some rain (or snow) at all times of the year. What they have in common is exposure to periodic outflows of very cold air masses from the arctic regions to the north, sending winter temperature far lower than the average for these latitudes. But these same regions, mostly on the eastern sides of the continents, get the benefit of warm ocean currents such as the Gulf Stream, bringing more rain and higher summer temperatures. The soils of these forests are mostly fertile and moisture-retentive, partly because they are young soils derived from glacial deposits. Climate and soils have not only given rise to this distinctive forest type but have encouraged its widespread destruction over many centuries, since farming has replaced forests. In large parts of China the destruction has been total except on steep terrain; less so in Europe, where the nobility have had their hunting reserves and forest production has also been valued more. In the USA settlers cleared vast amounts of deciduous forest for farming over the last four centuries.

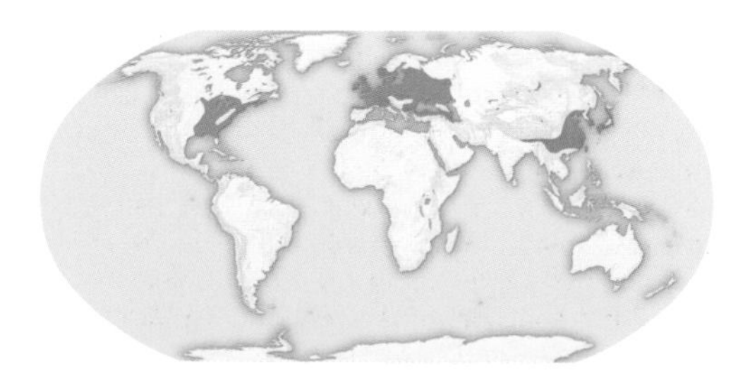

↑ **Fallen leaves and acorns** of the Valonia Oak (*Quercus ithaburensis*) in eastern Europe. The dead leaves of deciduous trees provide food for micro-organisms, which return their nutrients to the soil.

← **Deciduous forest** in Blackwater Falls State Park, West Virginia. Major tree species are likely to be White Oak, Tulip Tree, Sugar Maple and Yellow Birch. Fall color varies from species to species as does time of leaf-fall, as evidenced by the bare branches of some trees.

PRODUCTIVITY AND DIVERSITY

Deciduous forests are highly productive. With their broad, thin leaves angled to intercept maximum sunlight for photosynthesis, the trees are geared up to take maximum advantage of spring and summer warmth, storing away food energy in their leaves, branches, bark, wood and roots. The fall coloring coincides with the transfer of food out of the leaves and into the stem, so when leaf drop occurs they contain little more than the insoluble cellulose. Several genera of trees are common to all these deciduous forests. Most diverse are the oaks (*Quercus*) with at least 40 deciduous species in eastern North America, as many again in deciduous forests of China and Japan, fewer in Europe. Other prominent genera are the maples (*Acer*—most diverse in China), the elms (*Ulmus*), walnuts (*Juglans*), ashes (*Fraxinus*) and basswoods (*Tilia*).

Temperate broadleaf deciduous forest continued

Activity of the diverse lifeforms in temperate deciduous forest is determined by the seasons. In late spring after the leaves unfold, and through summer, there is often a superabundance of food. Light penetrates the trees' translucent foliage, allowing growth of lower plants on which browsing animals such as deer can feed. Different trees spread fruit production from the start of summer to mid-fall, so nut-eaters such as squirrels and some birds have a steady supply and can hoard for winter. And of course vast numbers of insects are supported by the trees. In some American oaks up to 30 species of leaf-miners (larvae of tiny moths), have been found in leaves of a single tree. The cycle of life, death and decay also speeds up in summer. Fallen leaves and branches are fed upon by microorganisms; in the case of wood, as in fallen trees and their stumps and roots, it is the fungi and certain insects that are first to utilize this resource. Fungi operate silently, spreading their mycelia through the dead wood, converting cellulose and lignin to higher-level nutrients including proteins. At some stage they gather energy to produce "fruiting bodies," the familiar mushrooms, toadstools, puffballs, bracket fungi and so on. These in turn are an important food source for many small fauna, some almost microscopic.

↑ **Male fallow deer** in a beech forest in southern Sweden, late summer. Well adapted to life in deciduous forest, fallow deer are not native to Europe but were long ago introduced from Middle Eastern mountains for hunting.

↗ **Eastern chipmunks** play an important role in the North American deciduous forest ecosystem by hoarding, and therefore distributing, the seeds of trees. They regularly sleep or nest in tree hollows.

← **A bracket fungus** on a fallen tree branch puts out numerous fruiting bodies. These are tough and leathery and may live for many years, a great contrast to the fruiting bodies of mushrooms and toadstools, which live only a few days.

→ **An acorn weevil** eats a hole in the acorn of an oak with its amazingly elongated snout. Reversing, it then lays an egg in the hole with its ovipositor; the hatched larva will eat the acorn from within. These weevils effectively compete with birds and squirrels as well as ground-dwelling animals for this food resource.

THE FOREST IN WINTER

In winter, life in the deciduous forest slows down drastically but is far from being at a standstill. Browsers such as deer must find food in the form of sugar-rich tree bark, twigs and buds, or scratch beneath snow or soil for mosses and roots. Tree-climbing animals such as squirrels also eat buds and consume their hoarded nuts. Many insects continue to feed on tree bark, wood or roots from the inside; many others go into a dormant pupal phase. Predatory fauna at all levels continue to seek their prey, except for those able to hibernate, like the black bear and most reptiles. Many birds evade winter by migrating to warmer climates, but others change their diet or depend on hoarded food. Some, such as owls, may find prey more easily in winter, in and beneath the trees' bare branches. Although growth of most trees stops completely in winter, some ground-dwelling flowering plants continue growth or flowering or fruiting, either into the beginning of winter or at its end before the spring thaw, and some of these are useful sources of food.

Temperate southern hemisphere broadleaf forest

This forest type is something of a mixed concept, defined more by what it is not, rather than what it is. It includes all the denser forests in temperate South America (which means only Chile and Argentina), Australia, New Zealand and the southern tip of Africa. Sparser tree-dominated communities in these same regions are classed as temperate woodland or savanna. Some of these southern hemisphere forests are really temperate rain forests, characterized by a closed canopy of few tree species with high occurrence of fern and moss epiphytes, sometimes with conifer crowns emerging, though not conifer-dominated. By contrast, the forest belt of temperate Australia is largely dominated by *Eucalyptus* species, with leathery, vertically oriented leaves. Epiphytes are rare or absent in these eucalypt forests and the relatively open canopy structure admits sufficient sun for a dense and diverse understory to develop. But eastern Australia also has many scattered patches of temperate rain forest, concentrated in the coolest, wettest places among mountains. Some of these, especially in Tasmania, are dominated by *Nothofagus* trees, and show strong similarities to the *Nothofagus* forests in New Zealand and South America—a result of the geologically quite recent splitting apart of the eastern part of the Gondwana supercontinent into these present-day landmasses.

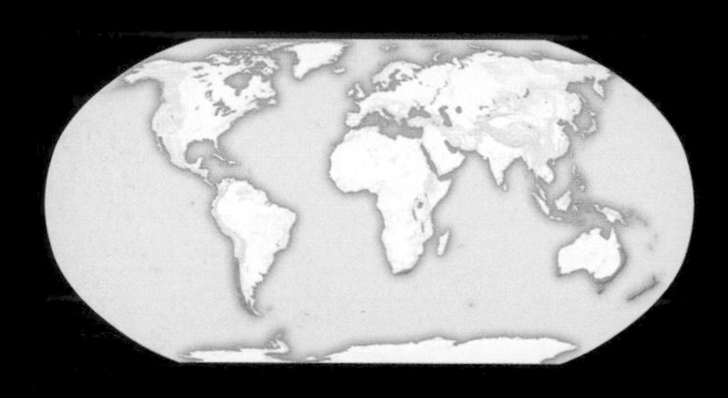

→ **This dwarfed forest** near the southern tip of South America consists largely of *Nothofagus antarctica*. Representing one extreme of southern hemisphere forests, tree size is limited by infertile soils and exposure to fierce winds.

↓ **The woody fruits** of this Gum tree release seeds after a wildfire, which fall into the ash below to germinate.

GONDWANAN LINKS

It is the Gondwanan element in temperate southern hemisphere forests that is especially interesting. Even in Africa, which split from the Gondwana supercontinent as far back as the Cretaceous period (approximately 100 million years ago) there are some trees of Gondwanan affinity, such as species of *Faurea* in the Protea family. But other genera of that family show much stronger links between Australia and South America, which were both joined to Antarctica as recently as 30 million years ago. As already noted, a major common element is the tree genus *Nothofagus* or southern beeches. The trees and other plants of Australian eucalyptus forests, in contrast, are largely native genera, though mostly with distant Gondwanan ancestors. Evolution over the last 40 million or so years has pushed them toward forms that are better adapted to lower rainfall and frequent wildfires.

Temperate southern hemisphere broadleaf forest continued

The various forms of life found in these southern hemisphere forests vary greatly depending on country and region. In the coolest temperate rain forests, with *Nothofagus* dominating, there are few larger mammals, birds or reptiles since food resources are scarce. Mist is a common feature of such forests. Smaller fauna are diverse but specialized and often hidden, including small rodents and marsupials, frogs and birds. Understory plants can be quite diverse, though concentrated in light gaps, and include many small-leafed, berry-bearing shrubs and scramblers. Insect life is diverse but also specialized. As in other forests of high-rainfall regions, fungi are common. It is with Australian eucalypt forests that some of the most abundant and diverse life-forms are associated. Apart from the many thousands of species of small trees, shrubs, grasses, ferns and other herbaceous plants, there is a wealth of fauna and fungi. Common larger mammals are the kangaroo and wallaby; the wombat, another forest-floor grazer; and the tree-climbing possum and koala, which live in the tree branches and eat the foliage.

→ **The koala (*Phascolarctos cinereus*)** is a marsupial whose sole diet consists of the leaves of the *Eucalyptus*, its digestive system tolerating the plant's toxic essential oils, though it prefers species with low oil and fiber content.

↘ **An adult cicada** emerging from its nymph stage, in which it may have spent as long as 17 years underground, feeding on roots. The adults sometimes producing a deafening symphony of continuous drumming and trilling.

↓ **A goanna climbing a eucalypt trunk.** These giant Australian lizards, often 6 feet (1.8 m) long, have a fierce appearance and, as recently discovered, a venomous bite, but are timid in the presence of humans.

LIFE AND REGENERATION

The Australian eucalypt forest is often noisy with the sound of bird calls. The birds range from the large wedge-tailed eagle, which can be found nesting in tall treetops, to tiny honey-eaters and scrub wrens. Most smaller birds are insectivorous, though nectar-feeders are also numerous, visiting a wide range of flowering trees and shrubs, including the eucalypts themselves. As in other forest types, insects and other small invertebrates are the most abundant fauna at the base of the food chain, though predatory and parasitic insects are, if anything, more diverse. All Australian eucalypt forests experience frequent wildfires and most of the plants, and even many of the animals, are adapted to survive these fires—in contrast to the moister temperate rain forests where the understory can rarely support a fire. Many plants sacrifice their foliage and sprout again from rootstocks or from buds buried beneath protective bark, others have woody fruits that release seeds post-fire to find a ready site for germination in the ash bed below.

↑ **The nocturnal sugar glider** feeds on the gum of acacia trees. It moves through the forest by means of two skin membranes which can be extended, like wings, to allow it to glide between trees.

← **The kakapo** is a near-flightless parrot with a very heavy body that lived in forest, scrub and grassland in New Zealand. Hunting and introduced feral animals rendered it extinct on the two large islands and it survives in very small numbers only on four smaller inshore islands.

Temperate woodlands

The term "woodland" is used in a somewhat narrower sense by plant ecologists than the way it is commonly understood in the English-speaking world, where it may be more or less equivalent to "forest." Ecologists use it to mean a rather lower, and sparser, kind of tree-dominated vegetation than one they would call forest, mostly with tree height of under about 50 feet (15 m) and the trees spaced apart often more than the average width of their crowns. The ample light reaching the ground in the woodland results in a well-developed understory of shrubs and herbaceous plants or of grass, depending on soil fertility and animal grazing. The term temperate woodlands refers to those that occur in the temperate zones, though not so close to the limits of those zones that they could be called either subarctic or subtropical. But the limits are also influenced by height above sea level, especially in the northern hemisphere, where increasing altitude may push both these boundaries southwards. Therefore, temperate woodlands may even occur south of the Tropic of Cancer, in regions such as the Mexican plateau and the highlands of Ethiopia. And there are also some extensive temperate woodland belts in some southern hemisphere countries, particularly Australia.

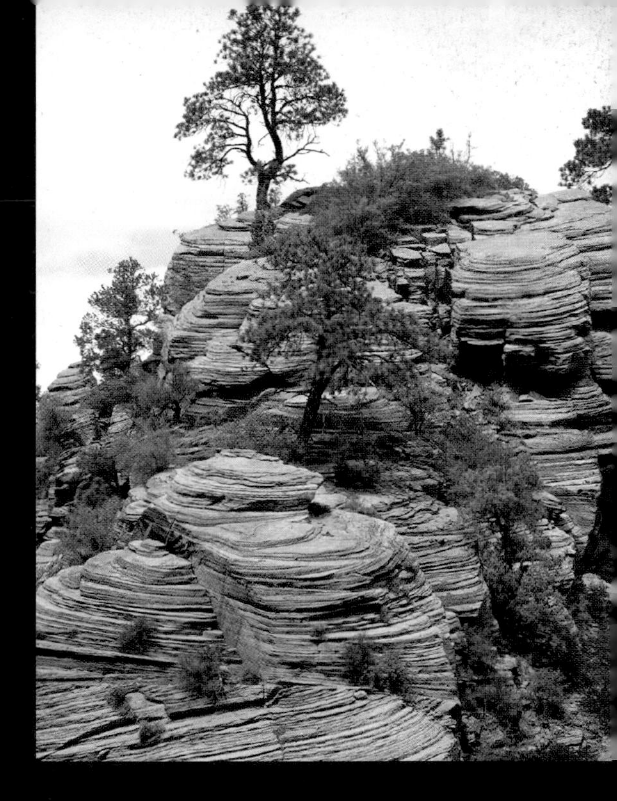

↑ **Sparse juniper-pinyon woodland** in Utah's Zion National Park is close to the higher and drier limits of this forest type. Pinyons are small pines (*Pinus edulis*, *P. monophylla*) with edible, nutritious seeds.

← **Sunset on Cathedral Rock, Arizona.** Riparian woodland of willows and poplars in the valley gives way on drier hill slopes to juniper-pinyon woodland.

CLIMATE AND THE TEMPERATE WOODLAND

Temperate woodlands are associated with drier and often warmer climates than the taller coniferous forests or deciduous broadleaf forests. In warmer temperate woodlands, the broadleaf trees are more commonly evergreen than deciduous while in the cooler types there is often a mix of conifers and both deciduous and evergreen broadleafs—such as on the extensive plateaus of western and south-central USA. In Europe the woodlands are concentrated in the Mediterranean regions, where summers are hot and dry and soils often shallow and exhausted; in some parts it may be that grazing and farming over many centuries has caused a change from forest to woodland. Another significant factor that can influence extent and structure of woodlands is fire. Periodic wildfires may remove grass competition and promote the mass regrowth of trees and shrubs in some woodland types.

Temperate woodlands continued

The major trees of temperate woodlands vary according to geographical region and climate zone. Around the Mediterranean, for example, the common species include pine, evergreen oak and some deciduous trees such as the maple *Acer monspessulanum*. In southwestern USA, evergreen oaks are also prominent, together with small trees of the heath family; in cooler and drier areas there is a transition to juniper-pinyon woodland. The open structure of temperate woodlands, with diverse shrub and herbaceous layers, provides habitat for a great diversity of fauna. Insects and birds are especially plentiful, sometimes sharing the nectar from the same flowering plants, though many other flowers are specialized for either insect or bird pollination; but the majority of insects feed on other plant parts such as leaves or bark—or on other insects—and most birds are insect-eaters or fruit or seed-eaters. There is an abundance and diversity of small mammals and reptiles, many of these also consuming insects and other invertebrates, and others preying on birds or other mammals and reptiles. Larger grazing and browsing mammals are mainly deer in the northern hemisphere, as well as wild pigs in Europe, and kangaroos and wallabies in Australia.

→ **The bobcat** occurs widely in the woodlands of North America as well as in other vegetation types. A medium-size member of the cat family, it preys mainly on rabbits and birds but will sometimes kill young deer or wild sheep.

↘ **The California condor** is the largest bird of prey in North America, with a wingspan of 9 feet (2.75 m). Once ranging widely over wooded hills and mountains of the southwestern states, it is now so rare as to be in danger of extinction.

↓ **Wild pigs** are one of the few large woodland mammals that have survived under heavy hunting pressure in Europe, perhaps because of their hardiness and intelligence.

LARGE MAMMALS OF THE WOODLANDS
Large predators are present in some temperate woodland regions, but are extinct in others, for example in much of Europe. In North America the mountain lion is the largest predatory mammal. Wolves once roamed the woodlands of Europe and North America, but have long since been exterminated. The Mediterranean woodlands have been much altered by millennia of human occupation. Apart from farming and grazing there has been so much pressure from hunting that most larger mammals and many birds are now rare or extinct there.

↓ **The Arizona mountain king snake** lives in drier woodland and desert regions of southern USA.

Tropical rain forest

The most biologically diverse of the Earth's biomes, tropical rain forest is found largely in the "wet tropics" where rainfall is high. At its highest development in the Amazon region, Central America, Central Africa and the Malay Archipelago, rain forest has a tall canopy that is almost "closed," meaning the tree crowns touch one another, leaving little space for sunlight to reach the ground below. There are often more than 100 tree species in one square mile of forest, and a huge diversity of life-forms, including ferns, mosses, orchids, bromeliads, shrubs, giant lianas hanging from the crowns, and other climbers clinging ivy-like to tree trunks. The forest floor may be so dim that only fungi are evident, but often there are sparse shrubs, palms and herbaceous plants as well. Many of the trees are heavily buttressed and strangler figs may be frequent. Tropical rain forest varies between the major landmasses in the types of dominant trees. And the wetter lowland rain forest grades into forests with increasingly more deciduous and smaller trees as regions of more seasonal rainfall are encountered; or into montane rain forest, also with lower trees of fewer species; on yet higher mountains this progresses to the enchanting cloud forests—also called mist-forest, mossy forest and elfin forest—where frequent cloud cover supports small, moss-draped trees perpetually dripping water.

STRUCTURE OF A RAIN FOREST

↑ **The complex structure** of a tropical rain forest, with an almost closed canopy, the presence of lianas, and regenerating trees in a hurricane devastated clearing.

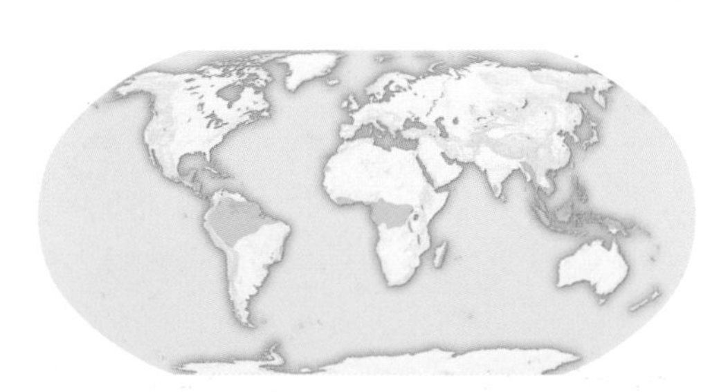

→ **Lowland rain forest** in the Danum Valley of Sabah, Borneo.

RAIN FOREST SUCCESSION

Regeneration of most rain forest trees is dependent on periodic disturbance of some kind, such as landslips, hurricanes or even the opening created when one over-aged giant crashes down, taking a swathe of other trees with it. In the temporary clearing or "light gap" there will rapidly sprout seedlings of fast-growing "pioneer" tree species. Also many other tree seedlings that have sat patiently in the deep shade making hardly any growth will rapidly take advantage of the light and newly released soil nutrients to quickly increase their height, the fastest eventually suppressing their slower rivals, and the longer lived species eventually replacing the pioneers. Every part of the forest passes through this succession over the course of several centuries.

Tropical rain forest continued

Most of the life associated with tropical rain forest is located in the canopy, among the forks, limbs and foliage of the tall trees. There is an abundance of epiphytes, lianas and creepers which contribute even more to the variety of foliage, flowers and fruits that provide shelter and sustenance to very diverse fauna. Insects and other arthropods exist in such huge variety that almost any biologist's collecting day in the canopy will yield species previously unknown to science. Many are close to the bottom of the food chain, nibbling at leaves or buds, "mining" through tissues of leaf, bark or fruit, sucking sap, sipping nectar from flowers or consuming decayed plant material that gathers in crevices. Insects are preyed upon by a large range of birds, reptiles, frogs and small mammals as well as by other insects. But there are at least as many bird and mammal species that feed directly on the plants. They include some larger mammals, in particular monkeys and apes, some species of which spend their whole lives in the canopy. And then there are a few predators at the very top of the food chain, such as eagles, some large snakes, and indeed some of the larger apes and monkeys who may quite happily catch and eat smaller monkeys.

↓ **The Toco Toucan** is the largest and one of the most colorful of the 40 or so species of toucan, all of which are native to the American tropics and subtropics. Dwellers in the rain forest canopy and mainly fruit-eaters, they have huge but lightweight beaks which may allow the birds to reach further along branches to get fruit.

← **A young Bornean orangutan** hangs from a rain forest liana. Orangutans are known only in the lowland rain forests of Borneo and Sumatra. The second-largest of the great apes after the gorillas, and the largest arboreal animal, they spend approximately 90 percent of their waking hours in the treetops, and construct sleeping platforms in the trees at night, by bending branches and twigs.

→ **The red-eyed leaf frog** (*Agalychnis callidryas*) is native to the tropical rain forests of Central America and, like other tree frogs, has adhesive disks on the bottom of its feet to aid it in climbing trees. It lays its eggs on the underside of the leaves of trees overhanging or near water. When the tadpoles hatch they drop from the tree into the water below to develop into frogs.

↘ **Found in Central and South America,** leafcutter ants forage in armies to cut and gather leaves which they bring back to the nest. The foliage is used for a fungus to grow on, which the ants cultivate in subterranean "gardens" for food.

AN INTERDEPENDENT RELATIONSHIP

Without nectar-feeding fauna many tree species would not have their flowers pollinated, resulting in a lack of seeds for reproduction. Moth-pollinated flowers tend to be white with a sweet evening scent and long tube into which only the moth's tongue can reach deep inside; bird-pollinated types are mostly red and visible from a distance, those attracting hummingbirds usually pendent on weak stalks; bat-pollinated flowers open at night with strong, often fruity or cheesy scent. Some fruits have relatively small, hard seeds that pass through the animal to be excreted later, often some distance away. The cassowaries of New Guinea and northeastern Australia, massive birds of the rain forest floor, can pass seeds up to 2 inches (5 cm) diameter through their gut. Other animals such as monkeys and apes chew off the fruit flesh and throw down the seed, while some fruit and seed eaters are simply messy and wasteful eaters, dropping a proportion of seeds or whole nuts while they feed. In these various ways seeds are dispersed to varying distances from the parent tree.

Tropical dry deciduous forest

People living outside the tropics are often unaware that deciduous trees are abundant over large areas of many tropical countries. In fact deciduous trees generally outnumber evergreens in tropical regions that have a long dry season, for example, in much of central India. The deciduous forest there was the original *jangal*, or jungle (though the word had a more general meaning of wooded or wasteland as opposed to cultivated). Losing its leaves in the tropical dry season is one way for a tree to avoid loss of water from its tissues. Many of the trees that lose their leaves do not become completely dormant, but expend some of their stored energy and water on the production of flowers and fruit, sometimes with spectacular displays of blossom on bare branches. Others wait for the wet season to flower and fruit. Tropical deciduous forest generally has a more open structure than rain forest and the trees are mostly smaller. There is often an abundance of climbers, but few epiphytes. Smaller shrubs and herbaceous plants are more frequent than in rain forest, but most of these are deciduous too, the herbaceous plants generally dying back in the dry season to underground storage organs such as tubers.

→ **Ceibo Trees (*Ceiba trichistandra*)** growing on a hillside in the arid, tropical forests of Manabi in Ecuador, South America.

↙ **Figs (*Ficus* spp.)** are a common feature of tropical dry deciduous forests, especially where rocky hills or outcrops provide a hold for their extensive aerial root systems. They have colonized some ruined temples in the tropics.

TREELIFE IN THE FOREST

Tropical deciduous forests tend to occupy the climatic belts that are bordered by rain forest on the wetter side and savanna on the drier. They are found mainly in inland regions of the continents or on larger islands such as Madagascar, Timor or Cuba. Very large expanses are found in central and southern Africa, southern Asia, South America and northern Australia. They often appear to be associated with poor or shallow soils or steeper topography, but that may be because the better sites they once grew on were cleared for agriculture. Many of the same plant families and genera that make up tropical rain forests are found in the tropical deciduous forest, though by a different set of species. Examples include the dipterocarp genera of Asia, *Dipterocarpus* and *Shorea*. Other plant groups come into their own in the tropical deciduous forest, for example the legume subfamilies Mimosoideae and Caesalpinioideae, with many but varied genera in all continents.

Tropical dry deciduous forest continued

The end of the tropical dry season is signaled by thunderstorms and afternoon downpours, and soon there is a burst of greenery from tree foliage, as well as the grass and herbage beneath. Grass- and leaf-eating animals such as deer and antelope thrive and breed. Large predators are likewise well provided for. Depending on location, the wet season may last from 2–6 months. Then the rains taper off and the dry season begins. This is often the hot season, with stifling heat and clear skies. Surface water all but disappears and the soil is parched; trees are leafless and the herbage beneath is dead. Larger herbivores must retreat to river banks, swamps and soaks and the large predators follow. But much of the fauna has ways of surviving the dry season. Smaller mammals dig up roots and tubers or eat the twigs and dormant buds of trees and shrubs; others eat insects. A major food resource in the dry season is the flowering trees, the nectar feasted upon by insects, birds and small mammals, especially bats. The whole forest is a complex system, with flowering and fruiting spread over the seasons, sustaining the fauna that the trees in turn require for pollination and dispersal.

↑ **A tarantula,** one of a large family of mainly tropical spiders that may often be found in deciduous dry forest. Ground-dwelling or tree-dwelling, they are nocturnal predators and include some of the largest and heaviest spiders, notably the bird-eating spiders, up to 4 inches (10 cm) long.

← **The brown lemur of Madagascar** inhabits an unusual kind of dry deciduous forest of small thorny trees. These primates are mainly herbivorous, eating tree leaves, fruit and flowers.

→ **A chameleon displays its camouflage ability,** changing color to blend with forest tree foliage. It is an ambush hunter of insects, using its long, sticky tongue which shoots out faster than the eye can see.

A COMPLEX FOOD CHAIN

The complexity of the food chain in a tropical dry deciduous forest is probably only exceeded by that in rain forest, with the extreme seasonality adding a further dimension. There is a huge diversity of insects and other small invertebrates, with many-layered prey–predator relationships even among these. Birds eat large numbers of insects, as do lizards, frogs and small mammals; these in turn are eaten by larger birds of prey and various other predators, such as snakes. A major insect group at the base of the food chain is the termites, feeding on tree roots and trunks, often present in astronomical numbers beneath the soil or under bark. They are important in the diet of many animals, for example armadillos and pangolins. Like many other insects their activity continues through the dry season, sustaining other fauna.

Savanna

The original savannas were in the Caribbean region and were probably grassy plains occurring in otherwise wooded regions, possibly on the large islands of Cuba and Hispaniola. Over several centuries the meaning of savanna has become more or less fixed to refer to flat or undulating country that is covered in grass or low shrubs, with trees widely scattered either singly or in copses with undergrowth. In the popular imagination, savannas are associated most strongly with Africa, as habitats of the "big game" animals—though often the image is of herds on an endless grass plain, which is at odds with ecologists' concept of savanna. As the map below indicates, savannas are found in tropical and subtropical regions where rainfall is strongly seasonal, usually with a short wet season and long dry season. The largest areas are in Africa, South America and Australia. Some of these areas border on deserts, with the savanna grading on the desert side into even sparser, lower trees and finally arid shrublands. In the other direction, that of higher rainfall, it may border on tropical dry deciduous forest or even rain forest. The change from forest to savanna may be influenced by soil type, by grazing animals or by incidence of fire, rather than simply by climate.

PLANT LIFE OF THE SAVANNA

Savanna trees are typically small to medium-size with spreading crowns. Certain plant families are represented to a disproportionate degree, in comparison to their frequency in other forest types. Foremost perhaps are the legumes, which have a deep-rooted habit able to extract moisture from deep in the subsoil. Typical of such legumes in all continents are the acacias, but they are joined by many other legume genera, especially in the Americas. Of major importance there also are species of *Prosopis* (mesquites or algarrobas) and *Mimosa* (the true mimosas). Another prominent family of tropical savanna trees in all continents is the Combretaceae, especially the genus *Terminalia* (which lacks a collective common name). Yet another plant group found on many savannas are the palms (family Arecaceae). In some savannas there is a great diversity of trees as well as of shrubs and ground-layer plants, most notably in the cerrado of central-eastern Brazil; but in some more arid savannas the diversity of trees and shrubs is low.

↑ **This diversity of vegetation types** on the African savanna supports the renowned diversity of large herbivores and carnivores, such as zebras, gazelles and wildebeests, in this famous big-game location.

← **Umbrella Acacias** (*Acacia tortilis*) frame a view over a corner of the Serengeti Plain, Tanzania. Savanna here alternates with treeless grassy plains and is also interrupted by lines of denser riparian forest along streams.

Savanna continued

Life-forms inhabiting savanna vary greatly beween each of the continents and large islands where it occurs. The large grazing mammals of the African savanna are so well known that they scarcely need mentioning, whereas American savannas have fewer and smaller kinds—mainly tapirs, deer, peccaries and rodents—and none of them are seen in large herds. In Australia the major grazers are the kangaroos and their smaller relatives. But in all tropical savannas some of the most important herbivores are the termites, usually only visible by the large mounds they construct—some over 20 feet (6 m) high. In the Brazilian cerrado, termites are commonly the most important herbivores in terms of quantity of plants consumed and they in turn are a food source for a wide range of predators—some, such as the American armadillos and giant anteater, and the Australian echidnas, are termite and ant specialists. Tree foliage is another major food source, exploited by browsers such as giraffes, monkeys and opossums, not to mention numerous leaf-eating insects, including ants. Insect life of all kinds is abundant and supports many birds, small reptiles and amphibians. Other birds feed on nectar, fruit or seeds, or prey upon smaller birds or lizards, as do many snakes.

↑ **The Jackalberry Tree** of Africa can be found growing on a termite mound as the termites aerate the soil, and the tree's roots in return protect the termites.

↑ **Patas monkeys,** widely distributed across west and central Africa, spend most of the day on the ground grazing on grass and foraging for food. They climb trees to sleep and as refuge from predators.

↗ **The giraffe's long neck and legs** gives it unique access to the treetops. Its long, sensitive lips and tongue allow it to nibble on tender young foliage and pods between the sharp spines of the thorny acacia.

→ **The Whistling Thorn (*Acacia drepanolobium*)** of the East African savanna develops these "pseudo-galls." Several ant species nest in them and swarm over the foliage, sipping from nectar glands on the leaves and discouraging browsers, such as giraffes—an example of plant-insect symbiosis.

← **Although one of the top predators** of the African savanna, the leopard is fearful of lions and requires trees for daytime refuge.

THE SAVANNA DRY SEASON

Savannas are often interspersed with rivers, smaller streams and swamps, and these serve as retreats for some of the fauna during the tropical dry season, when the swamps and riverbanks may contain the only green herbage. But just as in tropical deciduous forest, some savanna trees flower profusely in the dry season, providing nectar for insects, birds and bats; edible fruit and seeds may also support fauna in the dry season.

Mangroves

Mangroves are one of the most distinctive kinds of tree-dominated vegetation, defined by their occurrence below high-tide level on coasts and in estuaries and hence tolerating medium to high levels of salinity. Almost ubiquitous on tropical coasts, they extend well into the temperate zones in some parts of the world, mainly on moister eastern coasts that are washed by warm currents. Some of the farthest south mangroves are in South Africa, New Zealand and Argentina, and the farthest north mangroves occur in southern Japan, Louisiana, USA, and the top end of the Red Sea. Mangroves are most luxuriant and diverse where the seas are warmest and shallowest and there are large inputs of fresh water, silt and nutrients from rivers. The Ganges-Brahmaputra delta region of India and Bangladesh fulfills these requirements and supports one of the world's most extensive mangrove forests, though it is under some pressure from the dense human population. Conversely, on some steeper and drier tropical coasts, the mangrove band is very narrow and stunted, or mangroves may be confined to small stands in a few river mouths, for example, in Namibia and Peru.

MAJOR GENERA OF MANGROVES

The most important family of mangrove is the Rhizophoraceae, with the major genus being *Rhizophora* (Red Mangroves). Its species are stilt-rooted trees that often grow in the forefront of the mangrove belt, under attack from waves and ocean storms. Another significant genus is *Avicennia* (Gray Mangroves). These dominate mangroves at their cooler limits though they are also abundant in the tropics. They are not stilt-rooted but feature numerous pneumatophores, upward-growing roots that emerge above the mud and enable oxygen input to the root system. A feature of both these genera, and of some other mangroves, is their viviparous seeds which germinate on the parent tree and drop into the water to be carried by currents until stranded on mud, where they take root and grow.

→ **High tide submerges** the distinctive curved stilt roots of a Red Mangrove (*Rhizophora stylosa*), a species of the Pacific and Indian Oceans.

↓ **The massed pneumatophores** of the Crabapple Mangrove (*Sonneratia alba*) are exposed by the falling tide in this stand in Southeast Asia.

Mangroves continued

The mangrove forest is a truly unique environment. At the interface between land and sea and flushed twice-daily by the tides, it provides food and shelter to many organisms that regularly move in and out of it to adjacent environments, as well as other permanent inhabitants. The input of silt from freshwater streams, which the sea's waves and currents may redeposit along some shores, brings a rich supply of nutrients. The mangroves convert these nutrients into foliage which becomes a major base of the food chain. Although browsed by some animals and consumed by insects while still on the tree, it is when the leaves fall that they become an important food source. Vast numbers of animals, such as crabs and worms, dwell in the mud and consume the dead leaves and roots, while others, such as shrimp and fish, arrive on every tide. Apart from the mangrove leaves, there are algae that can also convert dissolved nutrients into food. In human economies one of the most vital roles of mangroves is their nurturing of fish populations. Not only do mangroves provide large quantities of food for fish, but the shallow waters are a breeding ground for some varieties of fish, where juvenile fish can be protected from ocean predators and are offered a range of food resources while growing.

↗ **The proboscis monkey,** so called due to its huge snout, lives in coastal areas of mangroves, swamps and riverine forests. It is both aboreal and amphibious, wading and swimming in water.

→ **The American brown pelican** nesting in mangroves at one extreme of its native range—the remote Galapagos Islands in the eastern Pacific.

← **Mudskippers** are remarkable small fish adapted by evolution to an amphibious lifestyle, hunting small fauna in the mangroves. They move by small jumps using enlarged pectoral fins and can breathe air so long as their mouth and skin remain wet.

THE MANGROVE FOOD CHAIN

Apart from animals that dwell in the mud or water beneath the mangrove trees, there is a wide range of fauna that moves through the branches and foliage, and amphibious animals that can move from water and mud to dry land. At the top of the food chain in much of the tropics are the crocodiles and alligators, preying on fish as well as large mammals and birds. A range of mammals from otters and water rats to monkeys, and even some of the big cats, hunt in the mangroves. Many kinds of birds also frequent mangroves, from eagles to small insectivorous types and nectar-feeders. Aquatic birds may nest in mangrove branches, and their eggs and nestlings are a common food source of snakes.

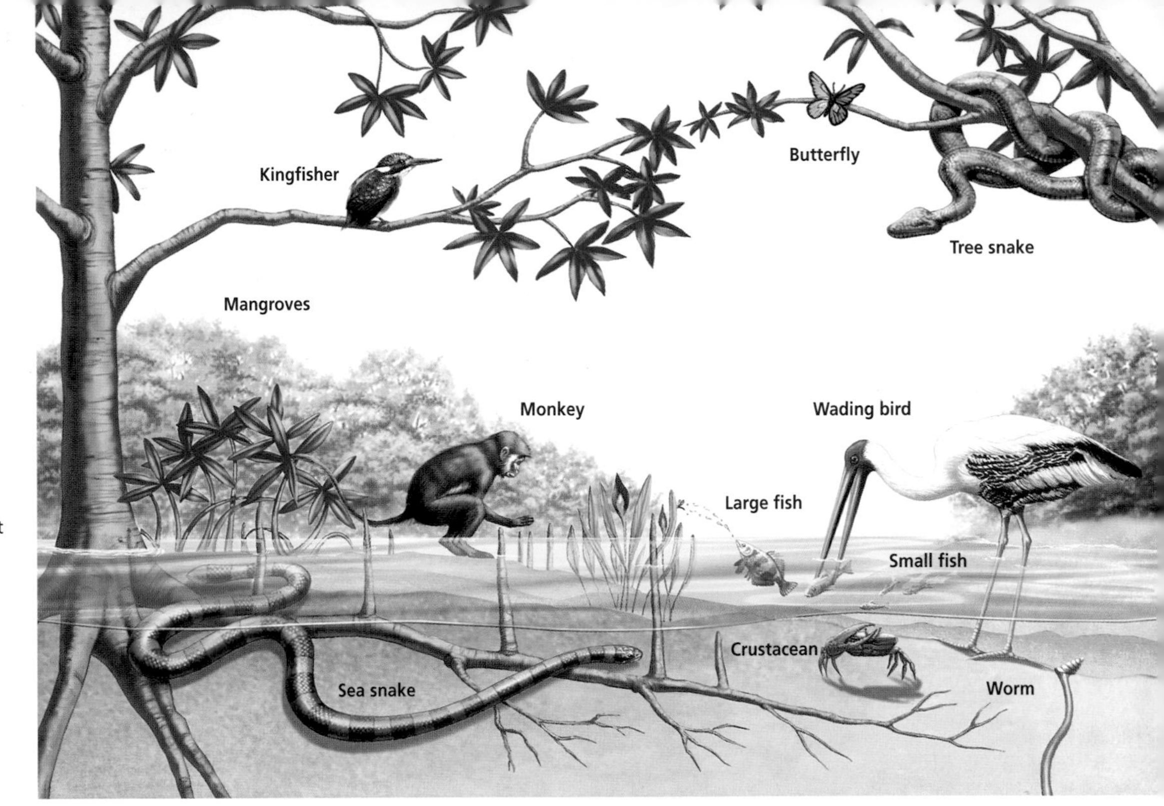

↓ **The giant mud crab** inhabits mangroves around the tropical and subtropical shores of the Indian and Pacific Oceans. It is carnivorous, feeding on smaller crabs, shellfish and worms as well as plant matter.

The urban landscape

Trees that are present in cities and their suburbs have two kinds of origin. Most common are planted trees, such as in parks and streets, or in private landowner's gardens. The second kind, absent from some cities but prominent in others, is remnant natural vegetation that has somehow survived as a city has grown up around it. In some cases survival has been a result of deliberate planning, but more often it has been due to the difficulty in building on certain types of land. Some of the most dramatic remnants of natural forest and scrub are found in cities with extremely steep hills; two outstanding examples being Rio de Janeiro and Hong Kong, while Los Angeles has its scrub-filled coastal canyons. Other kinds of remnant forest may be found in cities built on estuaries, with mangroves persisting. Many of the world's great cities have large areas of wooded parklands. The trees in some, even if originally planted, may grow and reproduce over centuries into a semblance of natural forest, as in parts of New York's Central Park and Paris's Bois de Boulogne. Trees in cities have some negative features. They untidily shed leaves, flowers and fruit on sidewalks, their roots interfere with building foundations, sidewalks and drains, and some trees have a habit of falling or shedding large limbs in storms, a potential risk to human life and cause of damage to vehicles. But trees are valued because they provide shade, shelter from wind, visual softening of the harsh cityscape, and often simply for the intrinsic beauty and interest of their form.

→ **A gray squirrel runs down a tree** in New York's Central Park. Squirrels are just one of hundreds of fauna species in this famous park, which occupies a large part of central Manhattan and was planted with hundreds of thousands of trees in the mid-nineteenth century.

↙ **Corcovardo mountain,** with its famous statue of Christ the Redeemer, rises out of Rio de Janeiro's Tijuca Forest to a height of 2,330 feet (710 m). Rain forest on the steepest slopes is remnant but much of the forest was replanted in the nineteenth century following earlier clearing. Visible behind are other steep, wooded hills entirely surrounded by city.

↘ **Suburban house blocks** in a new development in Tucson, Arizona, are each planted with identical mulberry trees, one landscape designer's idea of urban greenery. These trees are chosen for their ability to thrive in this quite arid climate.

INNER CITY WILDLIFE

Wherever trees are encouraged or permitted to survive in cities, wildlife will move in to exploit the many niches and food sources they provide—although the fauna, like the trees, are never as diverse as that of natural forest or woodland. However, there are many species that have adapted successfully to inner-urban living. English cities are famous for their populations of foxes, squirrels and badgers, and North American cities for squirrels, raccoons, skunks and coyotes. In Australian cities the leaf-eating, brush-tailed possum abounds in inner-urban parks. Around the world, city dwellers also receive pleasure from the wildlife attracted by urban trees. The morning chorus of bird call is balm to awakening commuters. And without trees there would be less appreciation of the seasons—the bursting of spring buds into tender new foliage and blossom, the summer's lush green shade, fall leaf coloring, and the tracery of bare winter branches.

Trees and the human world

Trees and the human world

From building materials, weapons and tools, to food, energy and pharmaceuticals, trees have been used in countless ways by humans for thousands of years. They have played an important role in the development of our societies, assisting us to defend or entertain ourselves, travel vast distances and even extend empires.

Totem, myth and symbol 224

Trees for timber 226

Homes and shelters 230

Getting from A to B 232

Across the water 234

War and peace 236

Decorative arts 238

Music and entertainment 240

Tools and utensils 242

Pulp and paper 244

Rubber, cork and fiber 246

Food and drink 248

Pharmaceuticals 250

Energy and fuel 252

Other uses 254

Ornament and amenity 256

The art of manipulation 260

Totem, myth and symbol

Trees have long been held in high regard by many cultures, and have been imbued with religious, magical or supernatural significance. The nymphs of Greek mythology inhabited trees and even became trees—the nymph Daphne was transformed into a laurel tree (*Laurus nobilis*) to escape the love-struck Apollo. In Norse culture, Yggdrasil was the World Tree, a giant ash that held Asgard, the home of the gods, in its branches. For the ancient Druids, oak trees were deified, and the word "druid" is derived from the Celtic word for wood. In religious texts, Buddha found enlightenment under the Bodhi Tree, while Adam and Eve were exiled from the Garden of Eden after eating from the Tree of Knowledge of Good and Evil. In northern Europe, long before the rise of Christianity, evergreens such as conifers and holly were revered as a sign of life. Bringing an evergreen branch indoors at the winter solstice, the shortest day of the year, brought good luck and the return of the sun. Today, this ancient custom is still celebrated with Christmas trees and wreaths, which are a mix of pagan symbolism and Christian beliefs.

→ **Totem poles** such as this cedar Tlingit totem from Ketchikan, Alaska, were made with local timber and natural paints. They could ward off bad spirits, recount a story or legend, or honor a birth or a death.

↘ **The modern Christmas tree** originated in West Germany. The custom was introduced to England in the nineteenth century by Queen Victoria's German-born husband, Prince Albert.

↓ **The Tree of Life** is a recurring theme around the world. In Malaysian mythology, the Tree of Life is said to have given birth to the first man and woman, when an upperworld vine impregnated an underworld tree.

↑ **An Assyrian stone carving** depicts a winged genie attending the sacred tree, which represents both the king, and the chief god of Assyria, Ashur.

→ **The Sacred Fig** (*Ficus religiosa*), or the Bodhi Tree, has great significance for Buddhists. Under just such a tree, Buddha meditated and achieved Nirvana.

THE TREE OF KNOWLEDGE

Many cultures have revered the tree for its supposed ability to elevate the human consciousness. The Lote Tree, in Islamic texts, is the farthermost boundary that one can progress in Heaven when attempting to approach God. However, the Lote Tree is not an actual tree, but rather a symbol for the highest point in religious knowledge that a human can attain. Many terms for knowledge or wisdom are derived from words for wood or tree. In ancient Scandinavian, the word for wood was *vid*. In Anglo-Saxon the word *witan* meant "mind" or "consciousness" and *witigia* was "wisdom." In the English language, related words are "wit," "witch" and "wizard" and, in German, *Witz* means "wit" or "joke."

Trees for timber

Trees from natural forests or plantations are logged to produce timber for building and other commercial purposes. Plantations are established in areas that have been cleared of original forest or reclaimed from agriculture. The principal plantation trees are conifers, or softwoods, and include Western Red Cedar (*Thuja plicata*), Monterey Pine (*Pinus radiata*) and Western Hemlock (*Tsuga heterophylla*). The quality of timber obtained from natural forests depends largely on those forests' age and location. A practice favored in managed forests is selection cutting, where only a proportion of trees is harvested. Forestry is a major industry in many developing countries, particularly in tropical and subtropical areas where rain forest hardwoods are logged. Removing trees from natural forests, particularly old-growth forests, is highly controversial because of such adverse environmental effects as soil erosion and the release of carbon dioxide into the atmosphere.

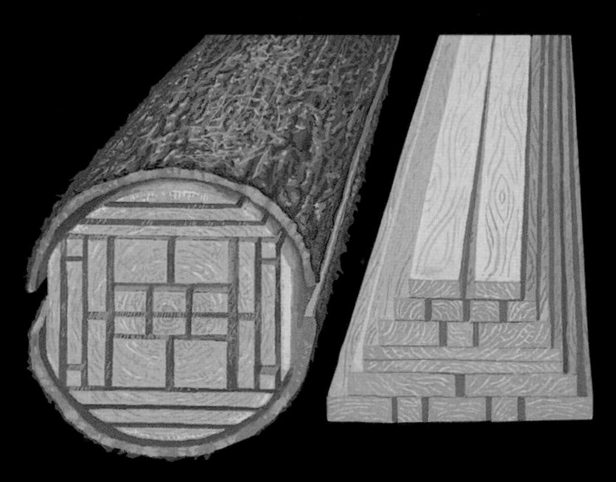

↑ **A sawmill uses a variety** of sawing patterns to cut a log, so that the resulting timber is of a uniform width and thickness. Lengths usually vary, however.

→ **These cedar planks** form vast stacks in a factory in Seattle, USA, in 1939. They were cut from trees felled by hand by lumberjacks.

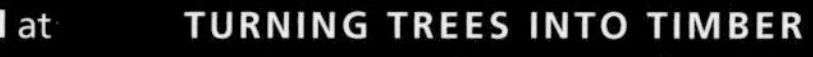

TURNING TREES INTO TIMBER

In recent years, logging and timber production have changed from a manual to a highly mechanized industry. Trees are felled, their branches removed, and then they are bucked—cut into logs. They are dragged out of the forest, then transported by road, rail or river to the sawmill. Here, the logs are sawn into planks of various sizes and shapes, which are standardized according to how the planks are to be used. Milled lumber is dried before being sold for construction, furniture and woodcraft. Some logs are cut into thin slices, called "veneer," to dress cheaper or less valuable timber or manufactured products. Other logs are turned into wood chips or pulp for the creation of paper or manufactured wood products.

← **Logs are washed** at the sawmill to remove debris such as stones and gravel. High-pressure water jets may also be used to remove the bark, so the logs are ready for processing. A sawmill can range from massive operations with computerized, laser-guided technology to small, portable mills for personal use.

← **Logs are loaded** for transport to the sawmill after they have been dragged out of the forest to the nearest road, railhead or waterway. In large operations, they are usually dragged by cables or pulled or shoveled by heavy machinery. In smaller operations, horses, mules or elephants may be used.

← **An alternative method** of processing logs is to turn them into wood chips or pulp, especially if the timber is low-quality. These products are used to create manufactured timber, such as particleboard and fiberboard, and paper.

→ **A tug pulls** softwood logs at Deception Pass in Washington State, USA. Where forests are near rivers or sheltered harbors, logs can be transported to sawmills in large, floating formations in a process known as "rafting."

Trees for timber continued

Depending on its source and strength, timber is divided into two categories: hardwood and softwood. Hardwood is derived from broadleaf tree species, such as eucalypts, oak, beech, and rain forest trees such as teak, while softwood is derived from conifers including pine, spruce and fir. Although there are many more hardwood species than there are softwood species, it is softwoods that form the bulk of commercial forestry operations. Hardwoods are generally slow-growing trees and produce wood that is more difficult to cut and work, while softwood trees usually grow much faster and produce wood that is ideal for many uses. Timbers are available in a range of colors and grains, the patterns being formed in the wood during the tree's growth and development of branches. The appearance of wood varies from species to species. Highly decorative or colored wood, such as rosewood or walnut, is favored for use in furniture and decoration. As many of these woods are rare or expensive they may be used as veneers rather than milled into timber. In this process, thin sheets of the veneer are applied as surfaces to other, cheaper timbers. Wood has always been favored for construction since it is readily available, can be easily worked and is strong and versatile. Although power tools are mainly used today for carpentry and woodworking, timber has traditionally been able to be cut and worked by hand. Old, seasoned wood is harder and less easy to work and shape than freshly cut timber.

↓ **Decorative timbers,** such as Jeffrey Pine (*Pinus jeffreyi*), are popular for furniture construction.

↓ **The knots in timber** are the impressions of branches or dormant buds from the living tree.

↓ **Plywood** is a manufactured timber comprised of several thin sheets of veneer stuck together.

← **Newly cut logs** are sprayed with water to preserve them and prevent them from drying out before they are taken to the sawmill for processing.

→ **When lumber is freshly cut, or green,** it has a high moisture content and is prone to warping and bending as it dries, making it unsuitable for most building and construction work. Lumber may be air-dried or kiln-dried. To ensure that cut lumber dries uniformly, it is often dried in large kilns, where the careful control of temperature, humidity and air circulation encourages slow, even drying. The air in the kiln is heated and circulated through the lumber by fans. The air absorbs the moisture from the heated lumber, the moisture condenses into water on a special evaporator coil and is drained out of the kiln. Most conventional high temperature kilns take approximately 10 hours to dry 1 inch (2.5 cm) of lumber.

WOOD-DRYING KILN

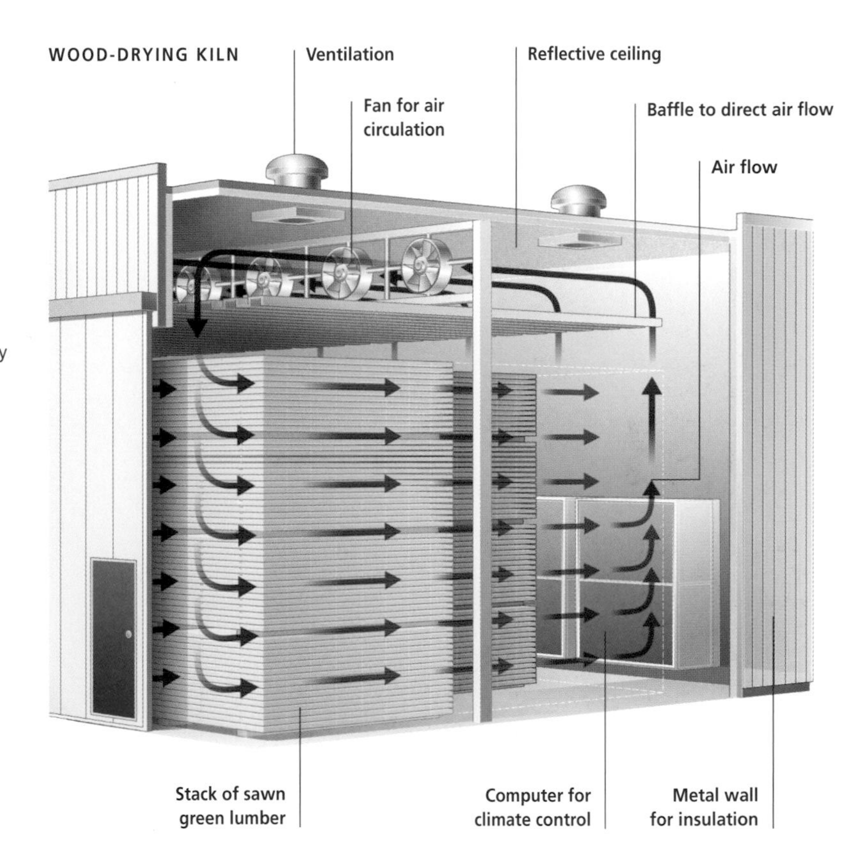

HARDWOOD VERSUS SOFTWOOD

Hardwood and softwood are terms describing the structure and origin of wood, rather than its strength. Hardwoods are derived from trees that produce flowers, known as angiosperms and include broadleaved trees such as oaks, eucalypts and tropical trees. Softwoods are derived from conifers such as pines and spruce. Conifers do not have true flowers for seed reproduction and are called gymnosperms. There are also cellular differences between softwoods and hardwoods. Softwoods have a simple cellular structure while hardwoods are more complex and varied.

↓ **This electromicrograph** shows the dense and complex structure of oak, a typical hardwood.

↓ **Softwoods, such as pine,** have a simpler and more uniform structure than hardwood.

↓ **The extremely lightweight** Balsa wood (*Ochroma pyramidale*) is actually a hardwood.

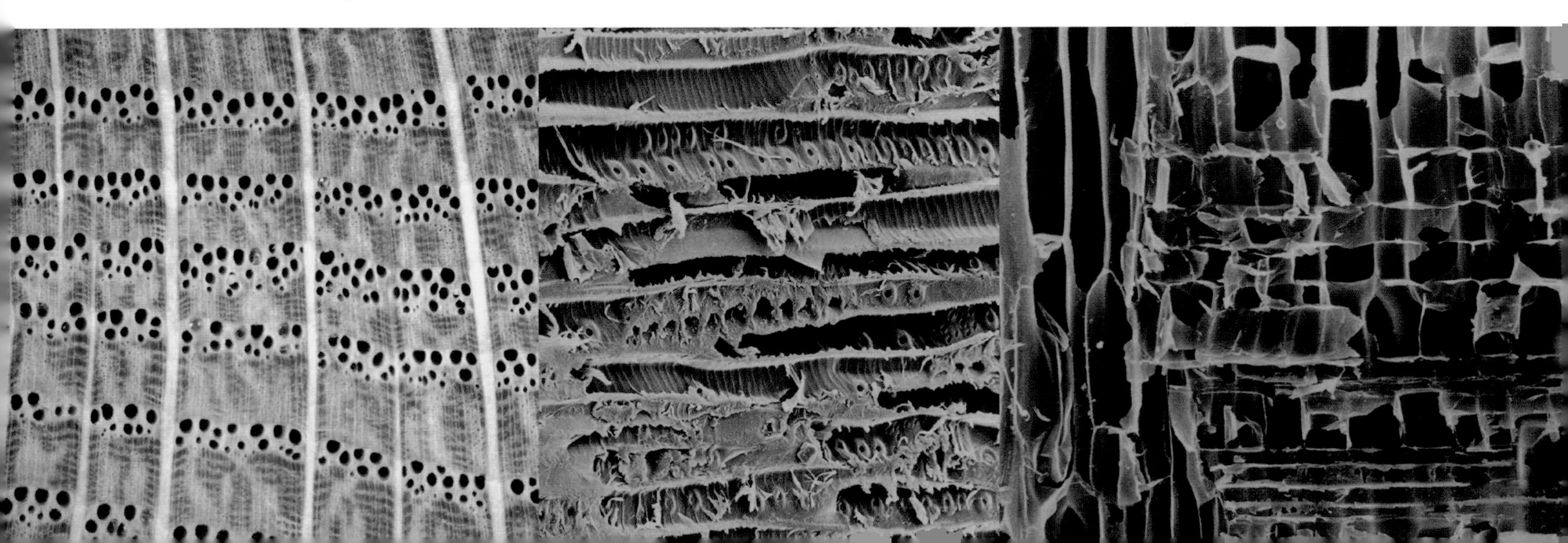

Homes and shelters

Throughout human history, trees and their by-products have been used to build structures as diverse as simple huts and one-room cabins, multistory houses, ornate and imposing churches and public buildings. Timbers used and methods of construction have depended on such factors as climate and other local environmental conditions and on the tools and technologies available. Where trees are plentiful, wood is usually favored over other construction materials. Even where stone, brick or animal hides are the principal components of a structure, wood is often required for frames, roofs, verandas or decorative features. Much modern housing consists of a timber frame that is clad in brick, stone, metal or concrete. Among the simplest human dwellings are bark huts—temporary shelters that are constructed of bark that has been stripped from trees. In tropical island communities where palm trees are plentiful, thatched palm fronds are widely used. In earlier times in parts of Europe, forests of oak and beech were widely exploited and distinctive building styles, such as the striking black-and-white half-timbered house, evolved. Long after the forests that provided their wood have disappeared, many centuries-old timber buildings survive intact. High-quality timber, despite being vulnerable to fire and weathering, is long-lasting and frequently reusable. Old timbers are often recycled in modern buildings.

← **Modern western houses** are still often constructed over a timber framework.

→ **Rows of timber** fishermen's houses—light, easy to transport and naturally insulated—are built into the cliff in the seaside town of Lysekil in Sweden.

↑ **Ornate timber paneling,** known as *boiserie*, was popular with French architects in the seventeenth and eighteenth centuries. In this process, timber was delicately carved, attached to a wall, door or piece of furniture and often gilded. These carvings would often take the form of pictures inside "frames."

→ **Stilt houses,** common on beaches and waterways in many southeast Asian tropical countries, are simple, economical and cool constructions made from timber and bamboo, with high-pitched thatched roofs for good ventilation.

← **The 22-domed** Church of the Transfiguration on Kizhi Island in northern Russia was built in 1722 completely from wood. It was constructed without using any nails or metal ties and even the joints are made from wood.

→ **The natural colors** and textures of pine, oak and many other timbers can be used to create striking home interiors. Timber wall paneling, ceiling beams and polished floorboards now feature prominently in many homes.

LIVING WITH TIMBER

Decorative woods may be featured in the construction and decoration of buildings. Wood can also modify the interior comfort of buildings by providing insulation against sound and weather. Wood detailing is often highly sought after and may add aesthetic value to a building. Original timber beams in a building dating from the fifteenth or sixteenth centuries, or parquetry floors in a twentieth-century home, are features that would be preserved and valued. Timber is also used to form decorative moldings and friezes. Doors and walls may be made of special timbers or clad in veneers. Floors may also be of wood and are often polished to show the beauty of the grain. Timber in walls and external detailing such as verandas, is usually protected from the elements by overhanging eaves and by coatings such as oil or paint, which protect and preserve the wood and may add to its decorative effects.

Getting from A to B

Wood has contributed significantly to the ways in which humans have discovered, explored and exploited their world, and has always played a vital role in the transportation of people and goods. Wooden sledges, pushed or dragged along the ground, were the first known vehicles. Around 3500 BC Sumerian craftsmen devised wheels made of wooden disks—contemporary stone carvings illustrate sledges with wheels attached. These wheeled wooden vehicles, pulled by animals or humans, could carry heavier loads, farther and faster than before, over all types of terrain. They opened up trade, enhanced construction and made possible the exploration of distant regions and, ultimately, the expansion of empires. From as early as 2000 BC, Egyptians and Hittites employed light timber chariots for both travel and warfare. Wooden carts and carriages were to remain the mainstay of land transportation until the twentieth century. Even early automobiles were made largely of wood, and timber is still used for decorative paneling on modern vehicles. Humans have even taken to the skies in timber craft. In 1903 the Wright brothers' Kitty Hawk—constructed of spruce, ash and muslin—made the first successful piloted flight. In 1947 industrialist Howard Hughes built the H4-Hercules aircraft almost completely of birch wood. It had the widest wing span of any aircraft, before or since.

↓ **Covered wooden bridges**, built from local timber in the nineteenth century, remain a feature of the northeastern states of the USA. The enclosed sides and roof protected the bridge structure from rot and wear.

← **Wooden sleds,** often pulled by dog teams, are light and move relatively easily over ice or through snow. They played an important role in the exploration and development of polar regions.

← **Wooden coaches** and carts, like this Turkish farm vehicle, were pulled by horses, oxen or goats. Before the development of the combustion engine, they were the main form of land transportation in most parts of the world.

→ **Hughes's "Spruce Goose"** was an airplane with a wingspan of 320 ft (97.5 m). Made largely of wood (birch), due to wartime metal shortages, it flew just once, in 1947.

CONQUERING NEW REGIONS

The existence of local timber contributed to the development of many remote regions, particularly in the eighteenth and nineteenth centuries. New settlers built simple log bridges that carried tracks and roads across watercourses and boggy ground. When construction methods improved, higher, flood-proof bridges replaced their crude predecessors. Timber poles carried telegraph lines and rapid communications between regions and countries. One of the longest was the 1990-mile (3200-km) Australian Overland Telegraph Line, built across the desert center of Australia between 1870 and 1872. Timber was also crucial in the development of railroads, which opened up inland areas on all the world's continents. The Trans-Siberian Railway, for example, permitted the easy movement of people, goods and produce through a vast, icebound region. Built between 1891 and 1916, it stretches 5770 miles (9288 km) from Moscow to Vladivostok.

→ **Train tracks** have been laid throughout the world using local timber railroad ties to support metal rails. They have allowed steam, diesel and electric trains to traverse vast areas with passengers and freight.

Across the water

Forests and the timber they provided allowed people from many ancient civilizations to venture onto the seas to discover new lands. The earliest floating craft were simple, wooden dugouts, bark canoes or rafts made from logs lashed together. These boats permitted their builders to fish and to travel to nearby coasts and islands. In prehistoric times small rafts and canoes undertook long sea journeys. The possibility of extended migratory voyages over long distances in prehistoric times was the subject of a study by Norwegian adventurer Thor Heyerdahl. In 1947 he built a traditional wooden raft, which he called *Kon-Tiki*, and sailed it from the Peruvian coast, across the Pacific Ocean to Polynesia. This journey of 3800 miles (7000 km) took 101 days to complete, and showed, at least in theory, that such a migration was possible. Among very early seafaring civilizations the Phoenicians were particularly successful. From their homelands in ancient Canaan (now Lebanon), Phoenicians traveled across the Mediterranean to establish trading posts throughout the region.

The mast and spars of the galleon were usually constructed from pine.

Hardwoods were used in construction of the hull and decking.

The keel was made from oak or mahogany.

Galleons such as Francis Drake's *Golden Hind* were among the most important sailing ships in history. Large galleons had several decks and could transport hundreds of sailors and passengers, plus enough cargo and food to last for several months.

↓ **The *Kon-Tiki*,** a balsa wood raft built following traditional methods, revolutionized ideas about sea migration in prehistoric times, when six men, led by Thor Heyerdahl, sailed from Peru to Polynesia.

↑ **Despite the widespread use of materials** such as fiber glass and aluminum in modern boat-building, many discerning owners still prefer craft made of wood. In Tasmania, Australia, for example, traditional timbers such as Huon Pine and King Billy Pine are still used to produce a wide range of small and middle-size craft. The decks of this privately owned luxury yacht are made from teak.

CONQUERING THE OCEANS

Distinctive styles and types of boats developed in different parts of the world. Native Americans built dugout canoes, which were often carved from single tree-trunks. Polynesian settlers in New Zealand traveled in canoes that Maori people call "wakas," and populations from Asian countries built wooden boats known as junks. In the Middle East, dhows sailed on seas and rivers. Traditional dhows can still be seen on the Nile in Egypt. In Venice, gondolas carried people and goods through its network of canals. They still ply that city's waterways, but mainly for the benefit of tourists. As boat-building skills developed, larger, safer wooden ships contributed to the expansion of trade, exploration and colonization. By the late fifteenth century, ships from Europe had reached the Americas and by the early sixteenth century, ships had circumnavigated the globe. They had traveled from Spain to Asia via the Indian Ocean, then crossed the Pacific and rounded Cape Horn to return to the Atlantic Ocean. As mapping and navigation methods improved, wooden sailing ships came into their own. During the seventeenth and eighteenth centuries they carried great numbers of people over long distances. Wooden ships ruled the seas until the early twentieth century when diesel and steam power usurped sail and metal displaced wood in ship construction. Wooden craft continue to be built—as luxury yachts or traditional fishing boats.

CONSTRUCTING A WOODEN DUGOUT CANOE

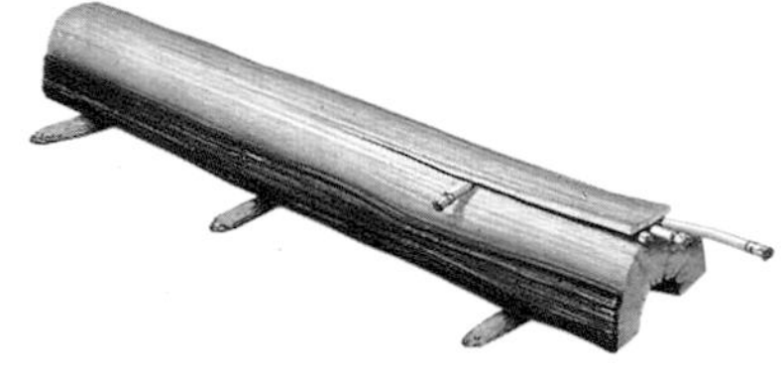

A cedar log was split lengthwise and then shaped using a stone tool.

The sides of the canoe were chipped away to reach the required thickness.

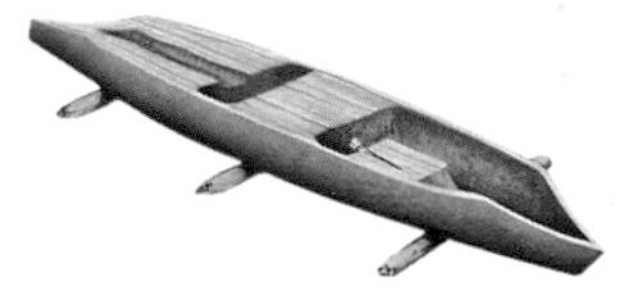

The inside was gradually hollowed out.

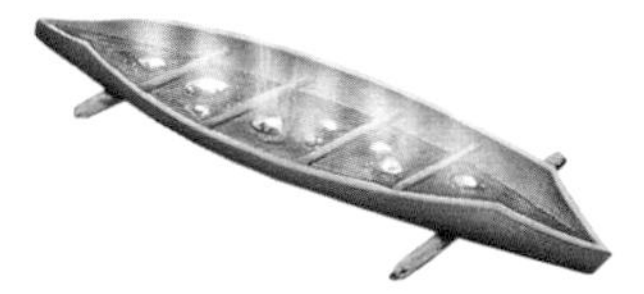

Water heated with hot rocks softened the wood.

The bow and stern pieces were attached and the hull sanded and decorated.

← **Native Americans** from northern California used the sea for hunting and for trade. They built one kind of canoe for the calmer waters of the bay and another, like this one at left, for the open ocean.

↓ **Dugout canoes,** such as this one from Papua New Guinea, are still made by traditional methods. Tools, such as adzes, are used to hollow out the logs or, in some cases, the interior of the log is burned out with fire.

War and peace

The first weapon was probably wooden—a heavy stick or lump of wood, picked up off the ground and wielded like a club. By the end of the Neolithic Period, humans had mastered the techniques needed to work wood and stone into spears, axes, flint daggers, arrows, bows, javelins, spear throwers, slings and throwing sticks. Some of the earliest known arrows, from Mesolithic Europe, were made of hazel (*Corylus avellana*) with flint or wooden tips. Examples include the blowpipe of the Dayak in Borneo or the boomerang of Australian Aborigines. These were hunting implements, but they were easily adapted into offensive or defensive weapons. The first known purpose-made weapon was a mace—a rock, later a lump of copper, hafted onto a wooden handle. To enhance their symbolic power, weapons were often decorated with carvings, precious metals, feathers or beads. Wood had the advantage of being abundant and easy to work. It was used for forts and stockades, shields and spears, battering rams and siege towers, chariots and gun carriages, catapults and crossbows. Even when metal, in the form of bullets and explosives, came to dominate war, wood was still used for gun and ax handles.

← **Metals** may have replaced wood in much weaponry, but wood is still favored for handles and supports because it offers good grip and transmits recoil well. Walnut, for example, is used for the butts of guns and rifles, while rare timbers such as ebony may be used for inlaid decoration.

← **Wooden shields,** such as these belonging to Masai warriors, have offered protection against arrows, spears and swords throughout history. Painted, carved or metal inlaid designs have also helped to identify individuals and their allegiances in the midst of battle.

→ **The Battle of Nájera** was fought between English and Franco-Spanish armies in 1367. The English, although outnumbered by more than two to one, used longbows to rout the Franco-Spanish, who were armed with cross-bows. In this fifteenth-century illustration, the English are on the left.

← **A Dayak man** from the nomadic Penan hunts with a blowpipe in the jungles of Borneo. The blowpipe is ironwood, the darts poison-tipped bamboo attached to a spongy timber. Many cultures have adapted their hunting weapons for the battlefield.

MACHINES OF WAR

Perhaps the most famous of wooden war machines is the Trojan Horse. Virgil's *Aeneid* tells of a huge horse made of pine planks which the Greeks used to smuggle soldiers into Troy and end a 10-year siege of the city. Wooden catapults were the first mechanical artillery. The Macedonians and Romans built huge timber siege towers. The lower stories had a battering ram and above were platforms for storming the city walls. Ancient Persians and Greeks used oar-powered timber warships as light, fast, ramming galleys. Vikings employed similar craft as raiding longships. By the seventeenth century sails had replaced oars and cannons were installed on timber ships.

→ **The trebuchet,** a more sophisticated and powerful version of the catapult, was used to bombard fortifications during medieval sieges. It had an arm that, with the release of tension, hurled stones, spears or other projectiles long distances. An invading army would normally use local timber to build and assemble a trebuchet at a siege site. With the invention of gunpowder the trebuchet finally went out of use.

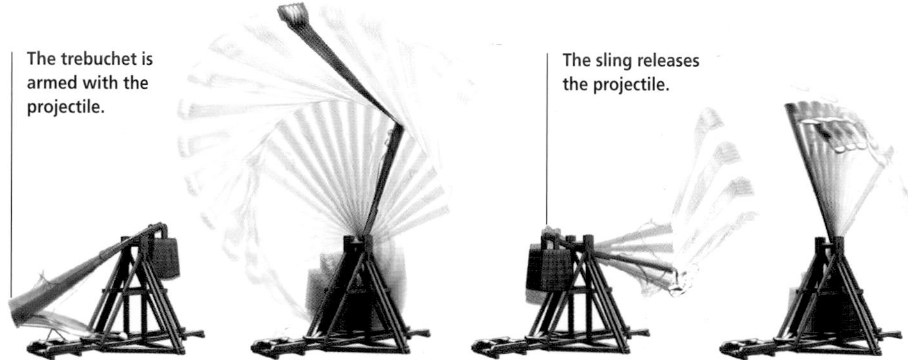

Decorative arts

Artists use timbers to create statuary and other artworks, as well as finely crafted furniture and utensils. Other traditional examples of fine woodcraft include figureheads on ships and facings on buildings. Wood is a versatile medium for an artist or craftsperson to work with. It may be carved, molded, whittled or shaped by bending or sanding to produce exquisite and highly detailed motifs and effects. Often it is the beauty of the wood itself that is most admired. Eighteenth-century English furniture makers such as Chippendale, Hepplewhite and Sheraton produced elegant but simple pieces that are still eagerly sought by collectors and connoisseurs. Timbers traditionally used for woodcraft, cabinetmaking and furniture include oak, mahogany, beech and pine. Woods that are highly colored or figured—those that display interesting grains or striking features when cut—are often used as decorative and contrasting inlays in tables and desktops. Popular inlay timbers include fruitwoods such as cherry, and rain forest timbers such as Ebony (*Diospyros ebenum*), which is almost black. This slow-growing timber, which is native to India, is becoming rare in the wild. To accentuate their natural grains and colors, timbers are rubbed with fine sandpaper then polished. Natural waxes such as beeswax, or stains, are used to enhance many timbers. French polishing—a painstaking process in which successive coats of shellac are applied and then rubbed back—imparts a mirror-like finish to a timber surface.

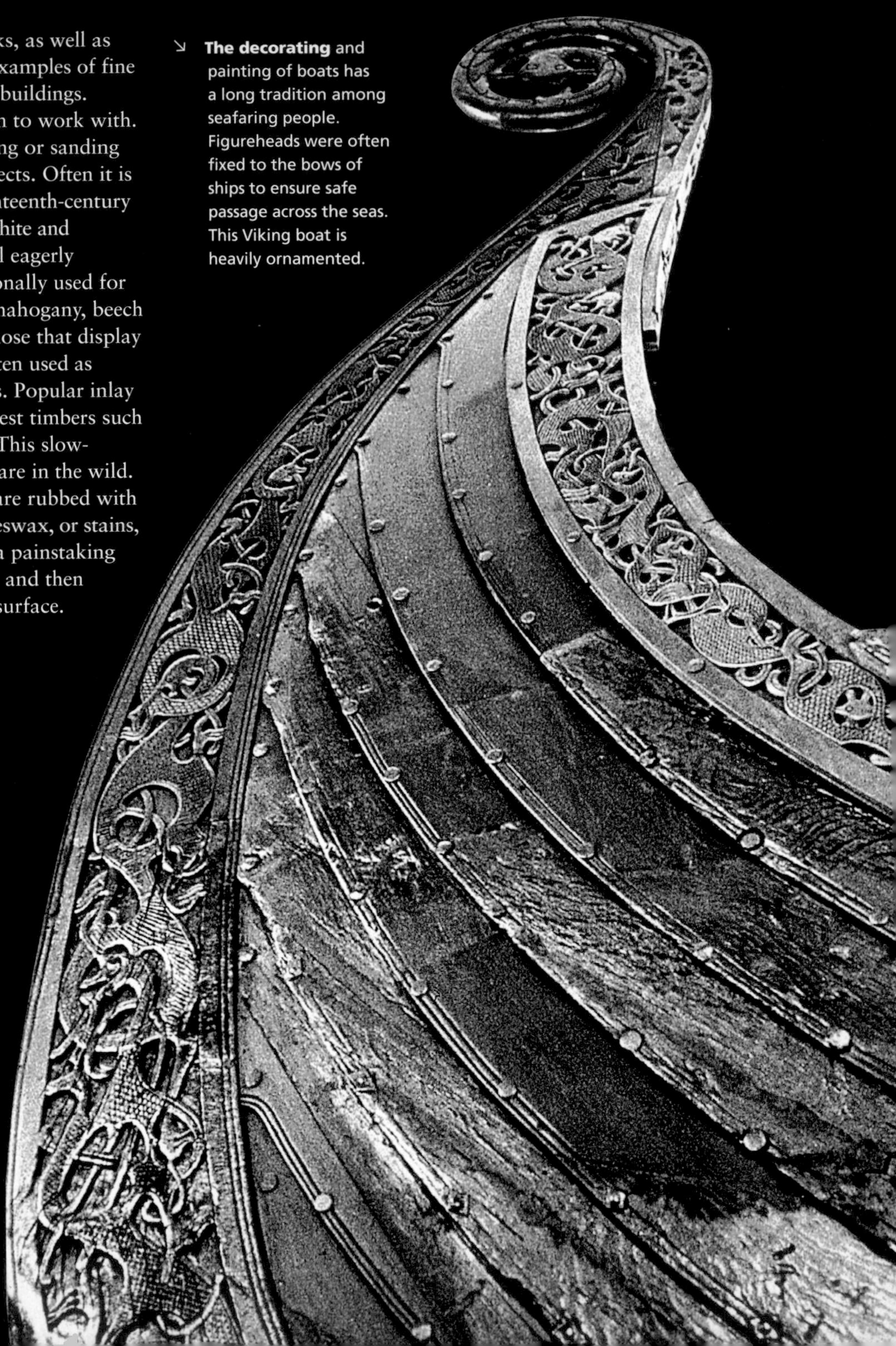

↘ **The decorating** and painting of boats has a long tradition among seafaring people. Figureheads were often fixed to the bows of ships to ensure safe passage across the seas. This Viking boat is heavily ornamented.

↑ **Amber** is a hard and ancient resin produced by conifers and valued as an ornament. It is popular for jewelry. Amber dates from many thousands of years ago and may contain the preserved bodies of insects or pieces of plants, some of which are now extinct.

RESINS AND FRAGRANT WOODS

Amber is a fossilized resin that was produced by ancient coniferous trees, including *Pinites succinifera*, which is known from the fossil and pollen records. Frequently, small organisms became trapped inside amber as it solidified. Amber is not a mineral but, because of its beauty, color, age and the quality of the encased organism, it is much valued as a gemstone. Most of the naturally occurring amber found today is between 30 and 90 million years old. The fossilized insects, spiders and other animals trapped inside are completely preserved with all of their soft tissues and provide invaluable information about aspects of prehistoric life. Other woods valued for jewelry include Blue Quandong (*Elaeocarpus angustifolius*), whose hard seeds are used in rosaries, leis and necklaces. Fragrant wood is also valued by craft and furniture makers. Sandalwood (*Santalum album*) is used for small timber ornaments and beads, as well as for making incense and as a lining for linen and storage chests. Camphor (*Cinnamomum comphora*) is another fragrant wood much in demand for chests as its camphor scent is said to deter moths and other insects.

↑ **Chippendale** is the term for various styles of furniture from the late eighteenth century, named after cabinetmaker Thomas Chippendale. Many other manufacturers of the period were inspired by his work. This carved mahogany chest of drawers was manufactured in Philadelphia, USA, between 1750 and 1760.

↖ **Wood turning** on a lathe is a popular woodworking craft often pursued by hobbyists. Wood turning utilizes attractive pieces of wood such as this chunk of Norfolk Island Pine (*Araucaria heterophylla*) which can be made into bowls and other decorative and utilitarian items.

← **The type of carvings** created by artisans reflect their beliefs and the wood available. This totemlike figure, carved from a tree trunk, reflects the Polynesian tradition of wood carving that stretches across the Pacific Ocean.

Music and entertainment

Musical instruments made from wood figure in cultures and civilizations in all parts of the world and throughout all ages, from ancient times to the present day. They range from implements as basic as African rain sticks, which are hollow wooden tubes filled with rattling loose beads or beans, to instruments as sophisticated as a Stradivarius violin or a modern grand piano. The natural acoustic properties of wood largely explain its particular appeal to makers of musical instruments. The didgeridoo, a traditional Australian Aboriginal wind instrument, and the sounding boards of stringed instruments such as violins, cellos and guitars, are based on wood's singular capacity to vibrate. Spruce is generally used for the soundboards of violins and the backs of violins and cellos are traditionally made from sycamore and maple. African blackwood—also known as rosewood—and ebony are the main constituents of wind instruments such as clarinets and oboes and of castanets. Traditionally, ebony has been used for the black keys of pianos.

↘ **Cricket bats** are made from willow wood. In the best wood the natural growth rings that run up and down the face of the bat are visible.

↓ **Simple flutes,** with several holes carved in hollow wood, are played by Quechua Indians in Bolivia. More complex flutes have keys and a much greater pitch range.

← **Simple drums,** made from wooden cylinders with a membrane stretched across the open top, feature in New Year celebrations in Beijing, China. These instruments have been made in the same way for thousands of years. As well as producing music, they are used for communicating over long distances.

→ **The eighteenth-century violin maker** Antonio Stradivari produced instruments of superlative quality. Scientific investigation of his violins suggests the wood he used gave his instruments their unique sound. It may have been very dense as a result of its slow growth.

← **Russian dolls,** called *matryoshka* or *babushka*, are cylindrically shaped groups of dolls made from wood that is colorfully and decoratively painted. Six or more dolls of ever-diminishing size nest inside one another.

→ **These rocking horses** are waiting to be painted and lacquered. Wooden rocking horses can be dated to seventeenth-century Europe. An early rocking horse was ridden by the future King Charles I of England when he was a child.

PLAYING GAMES

For thousands of years children have improvised simple toys and created games by using materials readily at hand. Generations of British children have fought mock battles with conkers—the seed capsules of a Horse Chestnut (*Aesculus hippocastanum*)—strung on sticks, or created simple dolls with clothes pegs wrapped in cloth. Some wooden games are enjoyed by both children and adults. Stick on Line, a traditional outdoor game that involves throwing marked wooden sticks, has been popular for centuries in Scandinavian countries. Other popular games that use wooden components include skittles, chess and checkers. Cricket bats are traditionally fashioned from willow, while strong, hard woods such as ash, maple and hickory are made into softball and baseball bats and golf clubs. Wood can also be crafted into toys and games that are objects of beauty in their own right.

Tools and utensils

The very earliest farmers probably made use of wood, dragging crudely sharpened sticks or branches along the ground to create a line or drill in which they could plant seeds or bulbs. As agriculture developed the simple wooden handheld stick evolved into a plow, which was pulled by domesticated animals such as oxen. Hollowed logs, or the fruit or seedpods from trees, were also used to carry water and store grains or were employed as receptacles in which to prepare or from which to eat foods. Empty coconut shells and the large fruits of the American Calabash tree (*Crescentia cujete*) served similar purposes. Before the advent of metal containers, a wide range of liquid and other agricultural products were stored in wooden barrels, made of a hardwood such as oak. Before they are bottled, many wines are still stored in oak barrels. Eating utensils have long been made from wood. It is estimated that 25 million trees are needed to supply the 45 billion pairs of disposable chopsticks that are used and discarded each year in China. The quality of wooden chopsticks varies enormously—from simple lengths of reconstituted unpainted wood fiber to fine, crafted utensils made from softwoods such as Japanese Yew (*Taxus cuspidata*).

WORKING WITH WOOD

Wood can be harvested green or it can be collected where it has fallen and fashioned with hand tools or fire to make hunting, agricultural or building implements. The earliest woodworking craftsmen used simple stone or bone tools to shape or split timber. With these basic materials they turned pieces of wood into rough hunting weapons such as spears, arrows or the boomerangs used by Australian Aborigines. Basic cutting implements consisted of pieces of stone—such as flint, which could easily be worked into a sharp edge—attached by strips of hide to hard wooden handles. Neanderthal and earlier peoples used these and similar hunting and cutting tools. When metals were discovered, wood provided the means to exploit them for productive use. Charcoal, formed from wood that was slowly burned, produced the heat required to melt and mold metals such as bronze, copper and tin and to forge iron.

↗ **As communities** settled and began to cultivate larger areas of land simple tools were replaced with more sophisticated, animal-drawn plows. This ox-drawn plow working a rice paddy in Bali, Indonesia, is fashioned from wood and has hardened metal tips.

→ **Simple woodworking tools** include chisels and planes to smooth and shape timber into any shape or product needed. In Siberia, the nomadic Nenets use hand-crafted timber sleds to transport their belongings

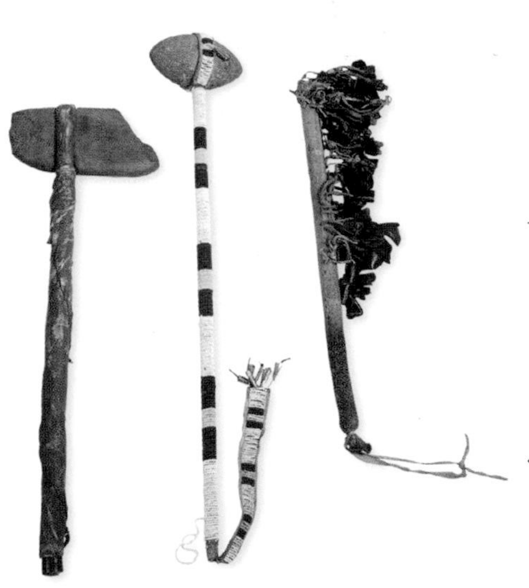

→ **Pencils made from wood** and graphite revolutionized communication from the mid-sixteenth century. Wood was used to encase brittle graphite to prevent it from breaking. In the earliest pencils, a groove was cut by hand to hold a slab of graphite. This was covered with a lid to completely encase the graphite. Favored wood for pencil making includes Californian incense cedar.

← **Early Native American tools** are typical of the most basic combinations of wood, stone or bone and animal hide to make simple tools for hunting, cutting and digging.

← **Barrel making** is an ancient skill that has changed little over the millennia. Most barrels are made of oak. Logs are split by hand and then aged in the open for several years. Machines cut the staves for barrels into required shapes and sizes. A cooper then manually assembles the staves inside several metal hoops. He heats the wood over an open fire until it is flexible enough for the remaining hoops to be fitted around them.

⇇ **Modern** tools and utensils are still largely made with wooden handles, which are comfortable to hold.

Pulp and paper

The manufacture of paper uses processes that were developed during the nineteenth century and which have been progressively refined ever since. The availability of cheap paper has had a significant impact on communications and other aspects of modern society. Trees provide most, but not all, of the raw material for paper and vast areas of managed forests exist to support people's seemingly insatiable hunger for this product. Trees selected for pulp are logged, debarked and then processed into fine wood chips. The chips are transported to pulp mills where, after being mixed with water and chemicals, they are subjected to pressure and reduced to pulp. This pulp is then refined, bleached, dyed and treated with additives until it forms a slush of fibers. The slush is spread over a moving wire screen to begin the drying process. As the slush dries, paper forms. Before mechanization was introduced, a manual process, which involved the use of a fiber slurry and a hand-held screen, or deckle, was used. The type and quality of paper produced, either manually or by machine, depends on the fibers used and the way the material is processed and dried. Wood pulp is also made into packaging materials. As well, there is an important manufacturing industry that recycles paper to make cardboard, other packaging materials, toilet paper and newsprint. Around 40 percent of paper is recycled to produce fiber. Wood chip is also a component of construction and craft materials such as MDF (medium density fiberboard), which is used for furniture and fittings. The location of paper mills is often a subject of controversy because of concerns about the environmental effects of contaminants such as chlorine bleach in the waste water produced by these mills.

↑ **Paper is formed from cellulose fibers,** seen here under a scanning electron microscope. Individual fibers are visible in handmade papers but in high-quality papers, the fibers bond together and cannot be seen.

↓ **As part of a growing industry,** waste paper products such as office paper and newspapers are collected and recycled for use in the manufacture of more paper, cardboard and newsprint.

EARLY PAPERMAKING

Though the word "paper" is derived from "papyrus," the processes used to make papyrus scrolls and fiber-based paper differ. To make a papyrus scroll, layers of papyrus were laminated together. Papyrus was used until the tenth century, when other papermaking techniques spread to the West from China. As early as 200 BC in China, techniques similar to those we now employ to produce paper from wood pulp were developed, using fibers from bamboo and mulberry trees. Papermaking also appeared in Mayan civilizations from about 500 BC. However, it was the skill and techniques of the Chinese that influenced papermaking around the world.

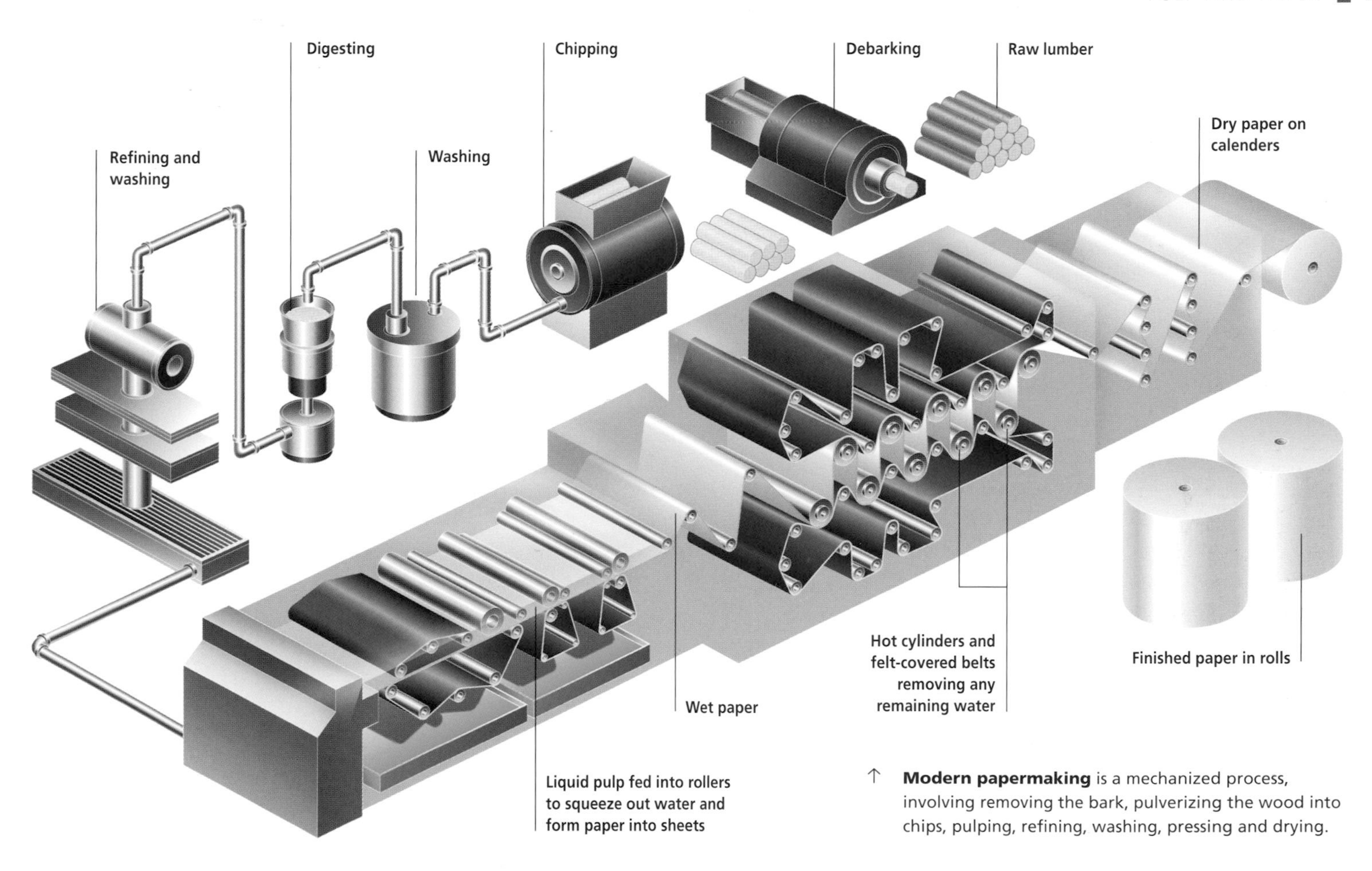

↑ **Modern papermaking** is a mechanized process, involving removing the bark, pulverizing the wood into chips, pulping, refining, washing, pressing and drying.

↓ **Wood fibers** are made into a pulp in the early stages of mechanized paper-making. The pulp is bleached then poured into flat sheets, pressed and dried into large rolls of paper. Technicians regularly check fibers to assess the quality.

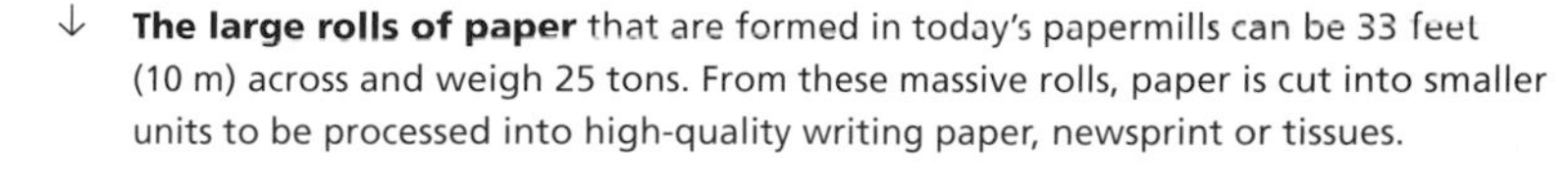

↓ **The large rolls of paper** that are formed in today's papermills can be 33 feet (10 m) across and weigh 25 tons. From these massive rolls, paper is cut into smaller units to be processed into high-quality writing paper, newsprint or tissues.

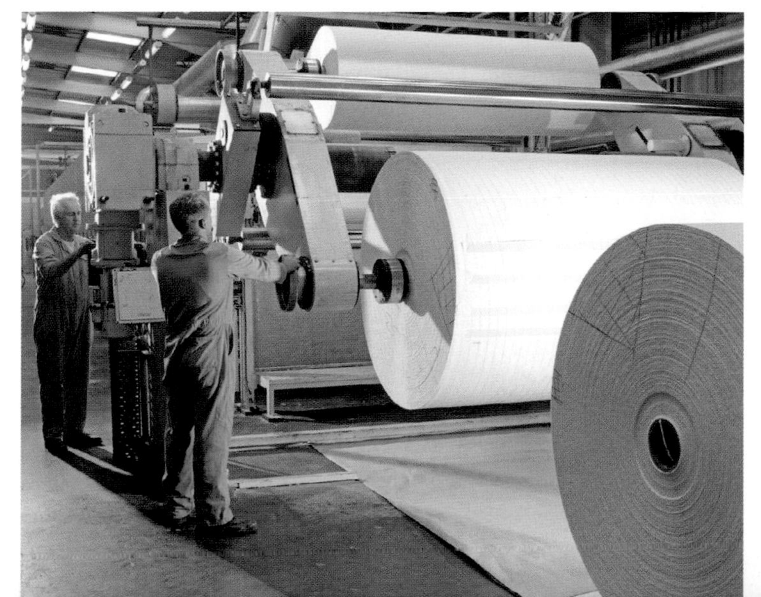

Rubber, cork and fiber

Trees have many economically important by-products. The milky sap of the Rubber Tree (*Hevea brasiliensis*) provides the raw ingredient for rubber. Until the development of synthetic products, the Rubber Tree provided all the world's rubber needs. Rubber manufacture begins with a process called "tapping," which allows latex to exude from the trunk. From the forest, latex is taken to factories that manufacture tires and other rubber products. Bark is another valuable by-product of trees. It is removed from lumber to be made into potting mixes and landscape mulches, although some bark is taken from trees that have not been felled. Bark from the Cork Oak (*Quercus suber*) provides cork for wine bottles, for buoying fishing nets and for floor coverings. Cork Oaks are native to the Mediterranean region; managed forests in Spain and Portugal provide 80 percent of the world's natural cork. Cork trees do not become productive until they are about 80 years old, at which stage bark can be stripped from them. The bark is harvested every nine years and the trees remain productive for more than 100 years. The recent trend to replace cork with screw tops in bottled wines is a serious threat to this centuries-old, sustainable industry. Oak trees are exploited for tanning agents and dyes as well as for their timber. Their acorns are sometimes used as fodder for pigs and other domestic animals and their flowers provide nectar for honeybees.

→ **A group of raffia workers** in Madagascar make souvenirs for tourists. Raffia is collected from a native Madagascan palm and now used around the world for weaving or in place of string.

↓ **Bark from Cork Oaks** is harvested at this plantation in Spain. The warm, sunny climate of the western Mediterranean region is ideal for the cultivation of cork oak trees.

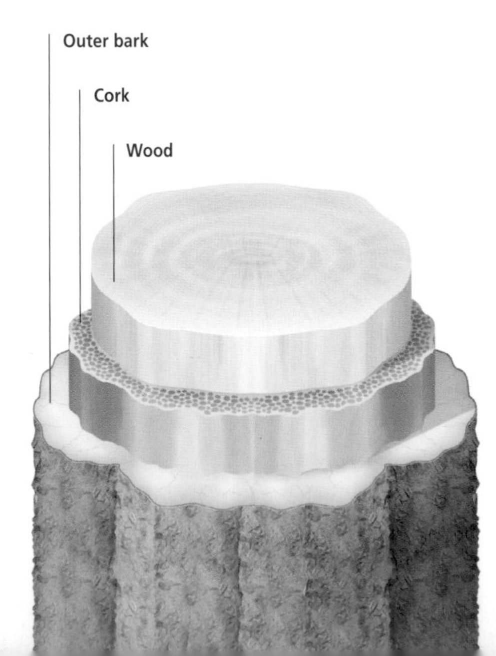

↓ **Cork is obtained** from the layer of thick, spongy bark of the Cork Oak, which is formed every time the rough outer bark of the tree is removed.

→ **Once harvested from wild trees,** 90 percent of latex used in rubber production is obtained from trees grown in commercial plantations in many tropical areas of the world, including Southeast Asia. The process of extracting latex from Rubber Trees is called tapping. The latex drips from a spiral cut in the bark and is captured in a large bucket. Tapping is often carried out at night when there is less risk that it will dry out.

→ **Rubber is used** to manufacture tires, mattresses, sheets, mats and hoses and is widely used in the mining and automotive industries.

← **Although the rubber tree** originated in the Amazon rain forest, most rubber plantations today are in Southeast Asia and tropical African countries such as Liberia. The raw product is taken to factories like this one on the Ivory Coast, for manufacture into rubber products. The latex is set in large trays and is then pressed and dried.

FIBER PRODUCTS

Raffia is harvested from the long leaves of the Raffia Palm (*Raphia farinifera*), which is native to Madagascar. The removal of the leaves does not destroy the trees. Raffia is used to make rope or is woven into bags, mats, hats and other textile products. Kapok, which has been used for centuries to stuff pillows and mattresses, is another tree by-product. It is derived from the silky down in the seed pods of several trees in the Bombacaceae family, including the Kapok Tree (*Ceiba pentandra*), which grows in parts of the East Indies, Africa and tropical America. The coarse fiber between the outer husk and inner shell of the Coconut Palm (*Cocos nucifera*) is called coir. It is used to weave floor coverings and, in recent years, has been composted as a peat substitute called cocopeat, which is widely used in potting mixes and as mulch.

Food and drink

Trees provide an enormous range of food, from fresh fruit and nuts to spices, beverages and oils. The edible sap of trees produces foods as diverse as maple syrup, which is gathered from Sugar Maples (*Acer saccarum*), and chicle, a traditional ingredient in chewing gum, obtained from the *Manilkara chicle* tree. Fruit trees grow in most parts of the world, from the tropics to cold-climate regions. Fruits that are harvested from trees include apples, plums, mangoes, coconuts, dates, oranges and other citrus fruits. Many agricultural economies are based around tree produce. Coconut and Oil Palms are vital for the livelihood of many Pacific Islanders, while dates are central to agriculture in desert communities. In Mediterranean countries, such as Greece, Italy and Spain, olives, cultivated both for their oil and fruit, are an important crop. Many of the world's major beverages also derive from tree crops. Coffee, produced from berries from the Coffee tree (*Coffea arabica*), a native of tropical Africa, is cultivated throughout tropical and subtropical regions and exported round the world. Early irrigation channels in Yemen indicate that coffee was already being grown in this part of the world in about 1350. Cocoa is extracted from the seeds, called beans, of the Cacao tree (*Theobroma cacao*), a large tropical tree from Mexico and Central America.

↑ **Spices,** such as the aromatic bark of the cinnamon tree, and the star-shaped fruit of star anise, add flavor to foods.

↓ **Citrus fruits,** including lemons, came originally from Asia but are now enjoyed around the world for their flavor and vitamin content.

SPICES AND FRUITS

Spices, many of which are produced from the fruit, bark or leaves of trees, played a vital role in the colonization of countries such as Indonesia and opened up trade routes between tropical regions and Old World centers in Europe. Indeed, the Dutch East India Company, which monopolized the spice trade, was the largest company in the world in the seventeenth century. Spices traded at this time included cinnamon, from the bark of the Cinnamon tree (*Cinnamomum verum*), mace and allspice. The history of fruit cultivation and exploitation reveals much about other aspects of human history. Breadfruit is linked to the story of slavery. In 1787 the British government sent William Bligh, as captain of the *Bounty*, to Tahiti to collect Breadfruit plants, which were seen as a potential source of food for slaves. Although this first voyage was interrupted by a mutiny, five years later 530 breadfruit plants arrived in the Caribbean. The story of the lemon is also linked to naval history. To overcome the problems caused during long sea voyages by scurvy, a debilitating disease caused by a lack of vitamin C, ships began to carry lemons and limes, which are rich sources of this vitamin.

↑ **Most of the world's nuts,** including almonds, Brazil nuts, hazelnuts and walnuts, are produced by trees. Nuts, which are eaten as snacks or pulverized into meal or flour for baking, are a protein-rich part of the diet of many countries around the world.

← **Olives trees** are long-lived and drought-resistant and can survive in poor soils. They provide a living and food source for millions of people, especially in Mediterranean countries.

→ **Breadfruit** (*Artocarpus altilis*) is a vital food in many tropical countries. The fruit gets its name from the fact that its taste when cooked is rather like that of freshly baked bread.

→ **Coffee production** is highly labor-intensive in regions such as South and Central America, where the beans are picked by hand. In areas, such as subtropical regions of Australia, where coffee is an emerging industry coffee is picked and processed by machine.

Pharmaceuticals

Much of what we know about the medicinal value of plants is traditional wisdom, handed down through many generations. Many such pharmaceuticals have been available commercially for more than a century. The bark of the evergreen Cinchona Tree (*Cinchona officinalis*) produces quinine, which has been used since the mid-seventeenth century to combat malaria. The ready availability of a powdered form of quinine in the mid-nineteenth century meant that Europeans could live safely in tropical malaria-affected regions. It gave a significant boost to the colonization of Africa and Asia that had begun more than a century earlier. Although quinine has been largely superseded by chlorinol, it is still used when strains of malaria prove resistant to the newer drug. The analgesic aspirin—acetylsalicylic acid—also came originally from a tree. Salicin, extracted from the White Willow (*Salix alba*), is the basic ingredient of aspirin. Another traditional medicine from a tree is ginkgo, from the conifer *Ginkgo biloba*. It is widely used in Chinese medicine to stimulate circulation and treat memory loss. Gum mastic from the Mastic tree (*Pistacia lentiscus*) is used in dentistry, while balm of Gilead, a resin from the Balsam Fir (*Abies balsamea*), has been used around the world for centuries as an antiseptic and healing agent. Two important antibacterial oils occur in native Australian trees. Eucalyptus oil is extracted from the Tasmanian Blue Gum (*Eucalyptus globulus*), while tea tree oil comes from paperbarks, including *Melaleuca alternifolia*.

→ **Antiseptic eucalyptus oil** is extracted from the fruits of the giant Tasmanian Blue Gum (*Eucalyptus globulus*), which soars to a height of 180 feet (55 m). In 1962 it was declared the floral emblem of Tasmania, Australia's southern island state.

→ **This is a computer model** of a molecule of the chemotherapy drug, taxol, used to treat ovarian, breast and other cancers. It comes from the bark of the Pacific Yew (*Taxus brevifolia*). The sphere-shaped atoms are color-coded to show carbon (pink), hydrogen (white), oxygen (green) and nitrogen (orange).

→ **The huge beanlike seeds** of the Moreton Bay Chestnut, or Black Bean (*Castanospermumn australe*), which are believed to hold substances that may assist in the fight against AIDS, are contained in pods up to 8 inches (20 cm) long. This evergreen tree is native to coastal areas of eastern Australia and some Pacific islands.

→ **In this magnified photograph** of White Willow (*Salix alba*) bark, the epithelial, or surface, cells within the bark are clearly visible. The bark contains salicin, from which salicylic acid, better known as aspirin, is derived.

Energy and fuel

Coal and peat had their beginnings in ancient plant life. Much of the coal that exists today began its formation during the Carboniferous period, between 280 and 345 million years ago. At that time, Earth was hot, humid and covered with vast forests and swamps. As trees died or dropped their leaves and branches into lakes and swamps, peat, a fuel that contains up to 90 percent water, was formed. Over time this peat became buried in sediment and compressed, or fossilized, into coal. This is why coal is referred to as a "fossil fuel." Most coal is mined either in underground or open-cut mines. It is used in coal-burning power plants to produce electricity and, because it is readily combustible and burns at high temperatures, is a traditional domestic heating fuel. Coal consists primarily of carbon but it can also contain sulfur, plant fossils and a number of other impurities. There are different grades, or ranks, of coals. The lowest grade—brown coal or lignite—contains the most impurities. These include hydrogen, oxygen and nitrogen, along with water and various volatile components. Large brown coal reserves are found in many parts of the world. Black, or bituminous, coal is a black or dark brown sedentary rock that fuels electric power stations. Coke is produced when bituminous coal is heated to very high temperatures out of contact with air. Anthracite, the highest grade of coal, has 95 percent carbon content. It is a metamorphic rock because the layers of decaying plant from which it formed have been subjected to pressure and extreme heat over a long period. Anthracite serves principally as heating fuel.

→ **Because of the heat they generate,** the power drills that that these underground coal miners are using are potential sources of danger. After millions of years forming underground, coal becomes hard and is highly flammable. To prevent it from combusting it must be wet when it is mined.

↙ **Since very early times,** humans have enjoyed the warmth and security of open wood-burning fires. The indiscriminate burning of wood can, however, lead to habitat destruction and other environmental damage.

↓ **When wood is burned in an oven,** such as in a wood-fired pizza oven, the oven temperature is controlled by regulating the amount of oxygen that is allowed to enter the fire.

WOOD AND CHARCOAL

Humans have been exploiting wood as a fuel for heating and cooking ever since they first learned how to make fire. As long as 5000 years ago people were using charcoal fires that were hot enough to forge metals such as iron ore and form them into weapons and tools. Wood is converted into charcoal when it is burned very slowly. Charcoal fires reach temperatures around 2000°F (1100°C), which is about twice the temperature of naturally burning wood. Charcoal making has a long tradition, especially in Europe. Here natural areas of forest with plants such as Hazel (*Corylus avellana*) have been continually coppiced for centuries to produce wood for burning in slow fires. To obtain the slow, hot burning needed to form charcoal, the fires are built in a covered pit where they are almost starved of air and are constantly tended to prevent them from going out.

PEAT, LIGNITE AND COAL

Peat forms from dead plant debris in waterlogged environments.

Lignite, or brown coal, is the lowest grade of coal.

Bituminous coal is a sedentary rock that fuels electric power stations.

Anthracite is the highest grade of coal and contains 95 percent carbon.

Other uses

The source of many food colorings and flavorings is found in trees. Annatto, which is used to color meats and stews, comes from the shrubby subtropical tree *Bixa orellana*, which is native to parts of Mexico and South America. The seeds that produce the coloring agent are contained in a red, soft-spined capsule. To derive the coloring, the seeds are steeped in hot oil or lard. A related food coloring, bixin, is obtained from the seed coat of the same tree. Bixin yields the yellow coloring found in cheese, margarine and butter, and the red color found in the skin of Edam cheese. While many synthetic products have replaced plant-derived products, the demand for bixin has increased in recent years as synthetic colorants have been banned. Many of the spruces (*Picea* spp.) produce oils and resins used for flavoring. Norway Spruce (*Picea abies*) is the source of spruce beer, as well as being popular as a cut Christmas tree. The new shoots in particular are used to make the beer. Norway Spruce is also used to bind cheeses, which develop a distinctive flavor from the spruce resin. Black Spruce (*Picea mariana*) is a basis for a beer known as Danzig spruce beer. The sap and young twigs are boiled with a sweetener, such as honey, molasses or maple sugar—another tree by-product—then fermented with yeast. This tree also produces spruce oil, which is used as a commercial flavoring. The leaves and twigs of Mugo Pine (*Pinus mugo* var. *pumilio*) produce pine needle oil, a commercial flavoring used in ice cream, sweets, beverages and gelatin.

DYES, THINNERS AND TANNINS

Other products from trees include skin dyes and preservatives. Henna, a hair and skin dye, comes from the small, tropical tree *Lawsonia inermis*. Wood creosote, derived from beech, maple and oak, is a disinfectant, laxative or cough treatment. In Japan it is in an anti-diarrhea medicine. When combined with petroleum-based products it forms black creosote, a timber preservative. Turpentine, used to thin paints and as a cleaner, is distilled from tree resins that contain turpenes, which are released by boiling. Trees harvested for wood turpentine include the Ponderosa Pine (*Pinus ponderosa*), Maritime Pine (*Pinus pinaster*), Loblolly Pine (*Pinus taeda*) and Aleppo Pine (*Pinus halepensis*). The residue is rosin, a varnish used on violin bows. Materials used in the tanning industry to preserve animal hides are also derived from trees such as oak, chestnut, willow and wattle. Tannin, a natural deterrent to fungus and bacteria, is found in high concentrations in bark, where it protects trees against wood rots and diseases.

The dye known as henna is obtained from the dried and ground leaves of the Mignonette Tree (*Lawsonia inermis*). Henna has long been used as a hair dye, as well as for decorations, such as the elaborate temporary tattoos that applied to the hands, forearms and feet of Hindu brides in India.

← **Bixin is a food coloring** that is obtained from the seeds of the tree *Bixa orellana*. The coloring is used for the red in the wax coating of cheeses, as well as for the yellow in cheese, butter and margarine.

→ **Slow-burning hickory chips,** or sawdust, impart a smoky flavor to fish and meats. The tree species used include Pignut (*Carya glabra*), Pecan (*Carya illinoensis*) and Shagback Hickory (*Carya ovata*), which is common in the southern United States.

← **Wood fibers, or cellulose,** can be used to form fabrics, cardboards and paper. Viscose, which is formed from cellulose, can be turned into fabrics such as rayon, and also into the wrapping paper called cellophane.

→ **The Freshwater Mangrove,** or *Barringtonia acutangula*, is used by some Australian Aboriginal tribes for fishing. The leaves, which exude a substance that stuns the fish, are thrown into the water then, after a period of time, the fish float to the surface and can be gathered.

Ornament and amenity

The first parks were established in Mesopotamia around the Tigris and Euphrates Rivers—an area now occupied by Iraq, parts of Syria, Turkey and Iran. Trees were planted on ziggurats (artificial hills) and in the famous Hanging Gardens of Babylon. In 1100 BC the ancient Assyrian ruler Tiglath-Pileser I created parks with avenues of trees that he took as trophies from countries he conquered in battle. In the more modern world, wherever space is available, trees dominate parks and gardens, arranged either in grand formal avenues such as in the seventeenth-century gardens of Versailles, or grouped in informal glades, such as in the naturalistic gardens of eighteenth-century Britain. In cool climates, deciduous trees are favored for parks, but in warmer regions, avenues may be planted with evergreens, including cypress, palm, eucalyptus or fig trees. One of the great parks of modern times is New York's Central Park, designed in the mid-nineteenth century by Frederick Olmstead and Calvert Vaux. Olmstead integrated the park and surrounding city through an innovative system of transverse roads, while incorporating the naturalistic charm of the site. Today's landscapers make use of modern technology to include mature trees in new or redeveloped parks, often transplanting trees from other locations to a park.

↑ **The magnificent Gardens of Versailles** in France were laid out for Louis XIV by the landscape gardeners Le Nôtre and Le Brun in the mid-seventeenth century and feature long, formal avenues of trees. Many of the trees that dominate the gardens today were replanted by Louis XVI in 1775.

← **In spring the area around the Jefferson Memorial** in Washington DC is awash with cherry blossom. In 1912 the Japanese Mayor Yukio Ozaki of Tokyo, donated 3000 'Yoshino' flowering cherry trees, which were planted near the Memorial around the Tidal Basin. Their blooming in early spring is celebrated each year with a festival attended by thousands of people, including visitors from around the world.

→ **The spectacular natural landscapes of Japan** are reflected in the plantings of palace and temple gardens. The gardens that surround the Enkoji Temple in Japan include Japanese Maples which are a blaze of color in the fall.

TREES FOR FLOWERS, FOLIAGE AND BARK

The earliest gardens included productive trees such as fruit trees but as communities became more settled, with more wealth and leisure time, trees became valued for their ornamental features. Large parks, roadside plantings and public and private gardens were created to feature trees for their seasonal beauty. Tropical gardens feature large spreading flowering trees such as jacaranda and bauhinia. In Asian gardens, trees are venerated for their rarity and age as well as for the seasonal beauty of their leaves, bark or flowers. Maples, ornamental plums and cherries, camellias and ginkgos are some of the trees featured in palace and temple gardens in China and Japan. Avenues or rows of trees are planted in large gardens, while small and home garden may include only one or two trees, often referred to as "specimen" or "feature" trees. These trees may be centered in a lawn or surrounded by flower beds.

THE LANDSCAPE GARDEN MOVEMENT

One of the greatest re-evaluations of the role of trees in gardens and parks occurred with the Picturesque or English landscape garden movement that reached its peak in late seventeenth- and eighteenth-century Britain, led by the great garden designs of Charles Bridgeman, William Kent and a gardener called Lancelot "Capability" Brown. Brown favored naturalistic plantings over formal. Grounds of large gardens such as Stowe in Buckinghamshire, were given a parklike look with groups of trees scattered in an apparently natural way. Brown also dug up formal parterre gardens, once popular around grand houses, replacing them with sweeping lawns, shrubberies and trees. The designs Brown executed were achieved largely through his and his clients' adoption of modern technology including new farming practices and the embracing of newly discovered trees from around the world. Apparently endless views from the house and nearby gardens to parks beyond where sheep, horses or cattle grazed were contrived through clever fencing, such as concealed ditches called ha-has and wire fences—as well as the artful positioning of trees. The landscape principles and gardens of Capability Brown opened up new directions in landscape design, which resonated around the world and continue to inspire modern plantings of trees today.

← **The beauty of the flowering Laburnum** (*Laburnum* x *watererii* 'Vossii') is accentuated through the art of pleaching. As the rows of closely planted trees grow, they are woven together to form a tunnel which is viewed at its peak in early spring when in bloom. This laburnum walk, underplanted with alliums, is at Barnsley House, Gloucestershire in England.

→ **In the early seventeenth century,** the Tuileries, with its long central axis grand avenues of trees, formal ponds and parterres, became the city of Paris' first major park. Today, intersected by the Champs Elysées, it remains one of the best examples of formal public landscape in the world.

← **The Jardin du Luxembourg** in Paris was built for Marie de Medici in the early seventeenth century, and combines elaborate parterres with shady avenues of elms and long gravel walks.

↓ **Palms are popular** trees for avenue plantings in warm and tropical cities including Los Angeles. Their tall, gray and imposing trunks form a colonnade overhung by giant green fronds. These Phoenix Palms appear to frame the world-famous Hollywood sign on the hills beyond.

The art of manipulation

Gardens are symbols of human control over the natural world. To maintain the size and shape of plants gardeners have developed various techniques. Pruning is for controlling a plant's size, shape and flower production. It can be the gentle removal of spent flowers to encourage new growth, the shaping of trees through topiary, or the restrictive annual removal of a full year's growth. This technique, often used with trees, is called pollarding. In some situations, trees are trained to grow flat against a wall, trellis or as a narrow screen. This is called espalier and is popular in small gardens. The art of bonsai, where plant growth is restricted by a combination of a small root system and regular pruning of roots, branches and leafy growths, represents the gardener's greatest control over nature. Branches and trunks of bonsai plants are trained and shaped using wires. Plants of all sizes including trees such as maples and pines are turned into bonsai specimens. Gardeners also create artificial climates and conditions in which to grow plants. These may be temperature-controlled glasshouses, which allow trees from climates as diverse as tropical rain forests and alpine mountains to be grown in one garden. In many of the world's botanic gardens plants from around the world are grown in glasshouses which provide warmth and humidity, for example the Tropical House in the Royal Botanic Gardens at Kew in London.

↓ **These gnarled and ancient Live Oaks** (*Quercus virginiana*) are native to southeastern USA. They were planted some 300 years ago to form a broad and impressive approach to this grand home in Louisiana. The trees are evergreen and have been carefully pruned to maintain the vista toward the house, while allowing the evergreen branches to intermingle to form a canopy. Another tree used around plantation homesteads and as street and park trees in the warm southern US climate, is the Bull Bay Magnolia (*Magnolia grandiflora*).

↑ **The pruning technique of topiary,** popular in formal garden styles, clips trees and shrubs into various elaborate and geometrical shapes. Yew trees are perfect specimens for this art form.

↑ **This normally large ginkgo** has been trained on a wall into a flat espalier. The tree's natural tendency to form a three-dimensional shape is controlled by removing all but selected horizontal branches.

↑ **Another way to manage the size** and shape of trees is to cut them back each year to remove the previous year's growth. This is known as pollarding.

BREEDING NEW PLANTS

As well as manipulating the way in which trees grow by pruning, training and climate control, trees and other plants have been manipulated for thousands of years through plant breeding and selection. Plants can form hybrids, which are natural or artificially induced crosses between plants of different species. Within a species some plants may also exhibit desirable garden or horticultural characteristics such as a smaller shape, larger flowers, tastier fruit or resistance to disease. Such plants are selected and multiplied by vegetative reproduction methods such as cuttings, grafting or division. A single plant may also form spontaneous variation such as a differently colored flower or variegated growth. Plant breeding can also be manipulated by exposure to radiation or chemical stimulation, which may encourage mutations which are seen as desirable.

→ **Bonsai,** an art that totally manipulates the growth and shape of plants, requires fine and patient work in shaping and handling plants that can be many hundreds of years old and of extreme value.

An indispensable resource

An indispensable resource

Trees are ubiquitous and are therefore often taken for granted. Many of us are unaware of their importance to our Earth and the part trees play in regulating our atmosphere and climate. Special programs are in progress in many areas to protect tree biodiversity and conserve this vital resource.

The importance of trees 266

The greenhouse effect 268

Trees and pollution 270

Biodiversity and conservation 272

Soil stabilization and erosion control 274

Flood protection and prevention 276

Deforestation 278

Reforestation 280

The importance of trees

Trees perform a vital role in the cycle of life on Earth. All life-forms depend on the ability of trees and other plants to photosynthesize—capture energy from sunlight and transform it into chemical energy to fuel growth. Growing trees are a food source for herbivorous animals plus a myriad of birds, insects and microflora. Carnivorous animals and organisms then feed on these plant-eaters. But not only are trees crucial to the food cycle. They are also vital to the carbon cycle. Animal life breathes in oxygen and breathes out carbon dioxide. Trees store the carbon dioxide and release oxygen back into the atmosphere through transpiration. However, carbon dioxide, known as "the greenhouse gas," is increasing in Earth's atmosphere, and this has been identified as a major cause of global warming. It has occurred during the past 150 years, mainly due to human activities. These include burning fossil fuels such as coal and gas, created by trees over millions of years, which create power for transportation, heating and industry. In addition, humans destroy carbon-storing forests in order to clear land for grazing and other forms of agriculture or to build houses and roads. Because trees store carbon as they grow, large-scale replanting of forests can begin to redress the carbon dioxide imbalance.

NATURE'S PROTECTORS

Trees are part of nature's huge recycling and protection system. As leaves, branches, flowers, fruits, seeds and even entire trees die and fall to the ground, they begin to decay. Insects, fungi and bacteria take part in breaking down organic matter. This releases nutrients and minerals back into the soil, thus making them available to other plants. Trees also recycle water, moving it from the ground back into the atmosphere. While trees grow, their roots help to stabilize the soil, reduce erosion and act as a buffer between land and sea. Mangroves, for example, help protect the coastline from the power of the ocean, while forest-clad mountains reduce run-off and soil and nutrient loss. On a large scale, forests can influence climate and weather patterns.

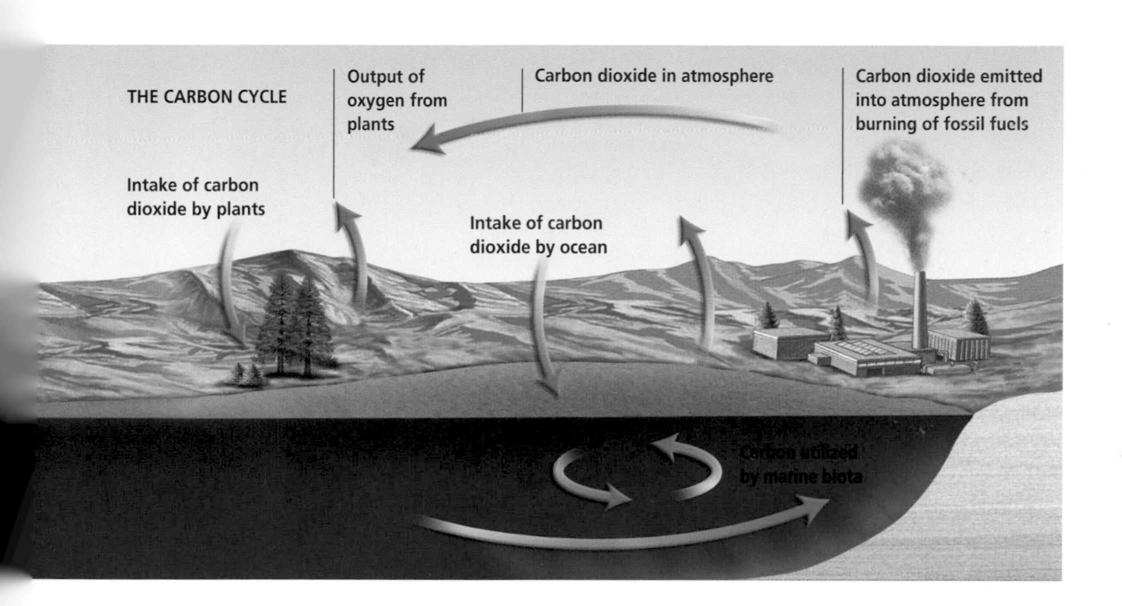

↑ **The three-toed sloth** lives in the forests of Central and South America and, just like vast numbers of other animals, it depends on trees for food and shelter. The sloth, known for its extremely slow metabolism, favors the rain forest species *Cecropia.*

↖ **Human activity** and animal respiration release carbon dioxide into the atmosphere. Some of this is taken up by growing plants and stored as carbon. The oceans also form a huge carbon reservoir. Cutting down or burning forests releases their stored carbon into the atmosphere.

← **Golden poplars** planted in intersecting rows act as a windbreak for orchards in the Neuquén Province in Argentina. They shelter the blossoms and fruit from harsh winds and help to ensure a good crop.

← **These crab apples** store energy as starch and sugars which will be consumed by birds and foraging animals. Many trees produce fruits that contain seeds, which are an important food source for animals, birds and insects.

↞ **Forests full of ancient trees** can store vast quantities of carbon over many centuries. Forests also affect the movement of air and water across Earth—they have a major influence on temperature, rainfall and local and global weather patterns.

The greenhouse effect

A natural biological process keeps a balance between atmospheric oxygen and carbon dioxide. As plants carry out photosynthesis they convert carbon dioxide into energy that is used in plant growth. An estimated 100 billion tons of carbon is captured annually by trees and other plants and stored in their woody tissue. This process of carbon storage is known as sequestration. As forests are cleared more carbon is released into the atmosphere. As human populations have exploded on Earth, developing vast industrial systems to support an urban-based lifestyle, more and more carbon has been released into the atmosphere. The main source of carbon dioxide release comes from the burning of fossil fuels to produce electricity. Increased carbon dioxide emissions are having a measurable effect on world climate. Carbon dioxide and some other gases trap heat in the atmosphere and are referred to as greenhouses gases because their presence mimics the warming created in an enclosed greenhouse. Increases in greenhouse gases in the atmosphere will lead to climate changes, particularly global warming. The rate and amount of temperature increase through climate change is not precisely known, but is projected to range from between 2 and 11.5°F (1.1–6.4°C) by the year 2100. Although carbon dioxide is the best known greenhouse gas, there are others, including methane, nitrous oxide and fluorinated gases such as hydrofluorocarbons. All have the potential to influence climate change.

↑ **Changes in climate,** caused by global warming, are likely to have adverse effects on the world's major forests which are already declining through clearing for agriculture, housing and logging.

→ **If the proposed climate warming occurs,** sea levels will rise over the next decade as ice caps at the poles, along with ice flows such as glaciers, melt.

NATURAL GREENHOUSE EFFECT

It is expected that forests and tree growth will be radically affected by increased levels of carbon dioxide and climate change, since forests are very sensitive to climatic variability. Changes in temperature, rainfall and atmospheric levels of carbon dioxide and other greenhouse gases all affect forest health. Some scientists suggest that increases in atmospheric carbon dioxide may actually benefit trees, leading to faster growth rates and greater sequestration of atmospheric carbon dioxide. However, these effects are likely to reduce as trees adjust to the increased levels of the gas over time.

THE GREENHOUSE EFFECT

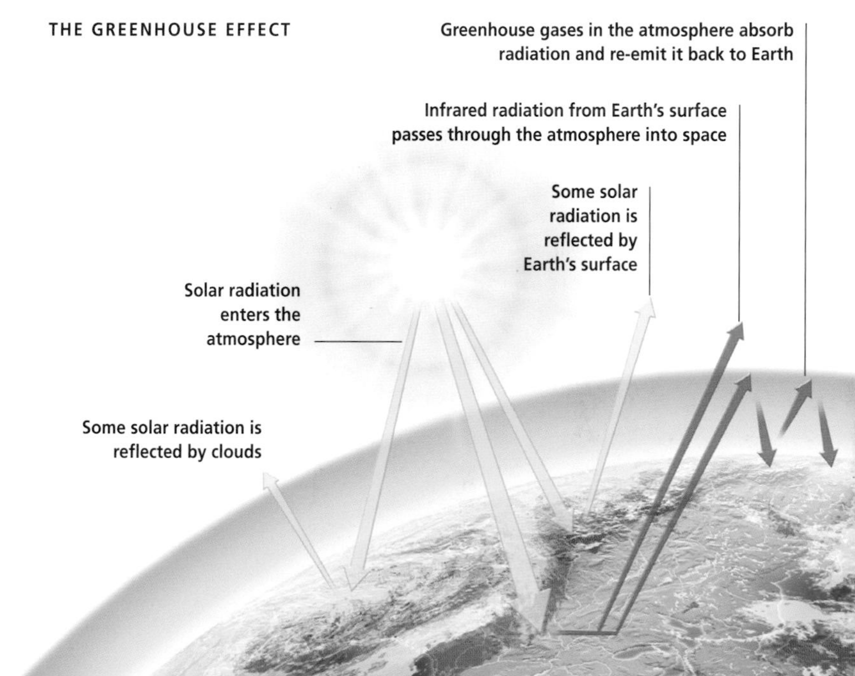

→ **Increased levels of greenhouse gases** in the atmosphere, such as carbon dioxide, absorb and reflect heat, leading to increased temperatures, as occurs within a greenhouse.

↓ **Burning fossil fuels like brown coal** releases carbon dioxide and other greenhouse gases. Much of the world's carbon is sequestered in coal deposits and in forests which act like large carbon sinks.

Trees and pollution

Along with extreme climate changes, trees are most at risk from pollution. Pollution is the contamination of air, water and soil and it can have a devastating impact on trees. Pollutants range from dusts to chemicals, such as sulfur. "Acid rain" is one of the most serious pollution threats for trees. It is produced when sulfur, as sulfur dioxide, and nitrogen, as nitrogen dioxide, are released into the atmosphere. These compounds combine with atmospheric moisture to form sulfuric or nitric acid. These acids can travel in the air and may reach areas far from their source before falling as acid rain. Atmospheric acids also contaminate snow, ice, fog and natural dust particles. When acid rain falls on leaves or needles it damages the foliage and also affects soil pH (the level of acidity). If it continues to fall unchecked, acid rain leads to defoliation, changes in soils, and ultimately the death of affected trees and other plants. In the northern hemisphere it has decimated vast tracts of natural conifer forests. The soil in many of these areas is already acidic and lacks the alkaline compounds necessary to neutralize the long-term effects of acid rain. In some cases pollution has caused trees to benefit since their insect and fungal predators are killed off. However, this may cause reduction in trees' immunity.

POLLEN POLLUTION

Trees themselves can be air pollutors. Many trees that reproduce by pollination are capable of releasing huge amounts of pollen to be carried on the breeze to the receptive flowers of other trees. But not all pollen reaches these flowers. Much of it remains airborne until it is washed from the air by rain or deposited on other surfaces. Spring, when many trees bloom and the air is filled with pollen, can be a difficult season for people susceptible to hay fever and asthma. Because most flowers bloom in the morning, pollen counts are usually highest before midday and are worst in dry conditions. The effects of these natural pollutants, which are seasonal and brief, are exacerbated by dust and by chemical contaminants released by agriculture and industry.

CAUSES AND EFFECTS OF ACID RAIN

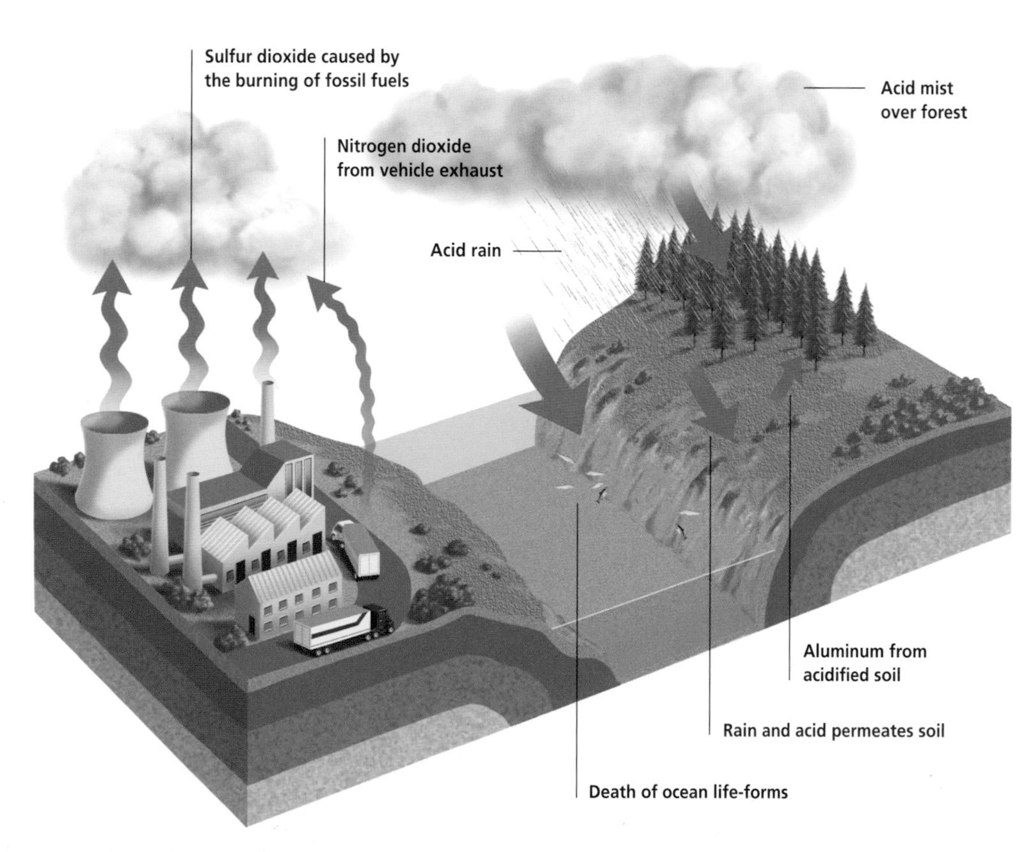

↑ **The world's worst nuclear accident** occurred at Chernobyl Nuclear Power Plant in the then Soviet Union in 1986. The resulting nuclear pollution destroyed or contaminated human, animal and plant life. Pine trees within a 4 square mile radius (10 km^2) turned brown and died from exposure to radiation. The dead forest became known as the "Red Forest." The worst affected area, now part of Belarus and the Ukraine, remains contaminated and is off limits to humans, however it has become a wildlife haven.

↖ **Mists and fogs** occur naturally over forest-clad valleys. Where the mists dissolve pollutants from vehicle exhausts and industrial emissions an unnatural form of fog called smog occurs.

← **Acid rain** is caused by pollutants such as sulfur dioxide and nitrogen dioxide combining with atmospheric moisture and then falling as precipitation. Acid rain damages plants and acidifies soil, and has caused conifer forests to die in many parts of the northern hemisphere.

⇇ **Oil refining** is just one of the industrial process that produces sulfur and other products that can form atmospheric acid compounds and eventually acid rain.

Biodiversity and conservation

The vast number and diversity of Earth's plants and animals is a testament to the success of life on Earth. Some 1.75 million species have been named and recorded, with countless more unknown or unnamed. By far the greatest variety occurs on land. Although terrestrial or land-based habitats account for only 29 percent of Earth's surface, they are occupied by around 1.5 million species. Some of the most diverse environments are found in the world's most extreme regions, such as the island of Madagascar—home to around 5 percent of the world's total plant species. Changes in habitat, most brought about by human activity such as land clearing and industrial development, are threatening biodiversity in all parts of the world. Plants play a pivotal role in assuring biodiversity in any ecosystem. It is estimated that for every plant that becomes extinct, as many as 30 other species, both animal and plant, may decline. To catalog plants under threat of extinction the "Red List" was compiled in 1997 by the IUCN (the World Conservation Union) and is continually updated.

HUMAN PROTECTION

Policies have been put in place around the world with the aim of protecting global biodiversity from further destruction. Many of these arose from the 1992 Earth Summit, where world leaders agreed on strategies to promote sustainable development, including the Convention on Biological Diversity. This convention aims to conserve biodiversity, to promote the sustainable use of plants and animals, and to ensure that benefits obtained from plant life are shared among the world's people. In 2002, a Global Strategy for Plant Conservation set 16 targets to protect threatened areas of biodiversity and to conserve plant species through seed and tissue banks, and through cultivation. These targets are to be achieved by 2010.

↑ **In a bid to protect biodiversity** seeds and tissues from plants around the world are collected and stored in climate-controlled seed banks. The largest resides in London at the Millennium Seed Bank at the Royal Botanic Gardens, Kew Gardens, England.

← **Natural climate extremes** can adversely affect both natural and urban landscapes. These street trees in Beijing, China, have been wrapped to protect them from frost and storm damage over winter.

↓ **Strip cutting** is a sustainable method of cutting timber on hilly terrain. The mature trees on the top of the hill are left untouched to provide a source of seed for regeneration, and the leaf litter naturally falls downhill to create nutrients for the growing forest.

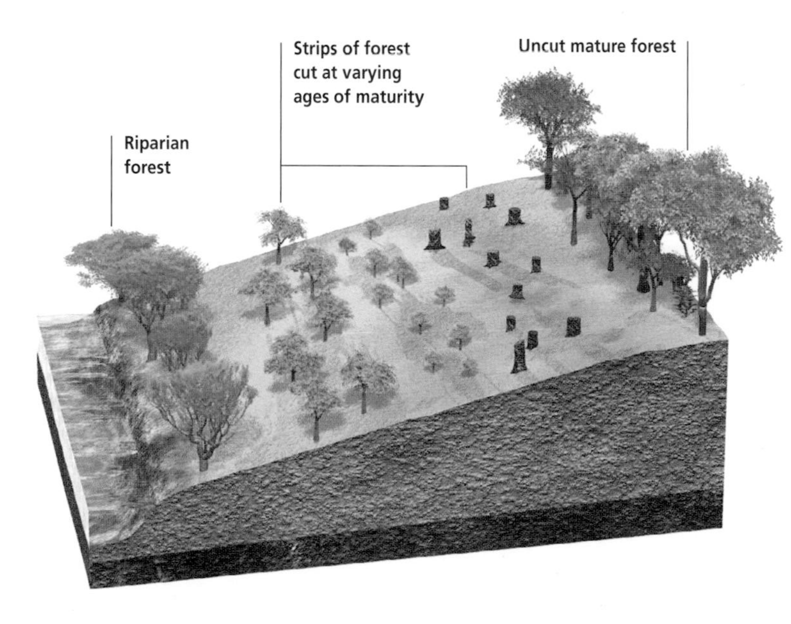

↑ **Biodiversity,** such as found in this Australian rain forest at Mossman, Queensland, is under threat by human-induced clearing and climate change.

→ **The Wollemi Pine** in the family Araucariaceae was discovered in Australia in 1994. One of the oldest and rarest plant species in the world, it has been identified from fossil records dating from 90 million years ago.

Soil stabilization and erosion control

Soil erosion by wind and rain has long-term consequences for a variety of ecosystems. Areas that are cleared of vegetation through practices such as deforestation are highly vulnerable to erosion from both wind and rain. Once the nutrient-rich topsoil is removed from an area it is less capable of supporting a healthy community of plants. As well as causing damage to the areas where the soil is taken from, soil erosion has a detrimental effect on other areas. Wind-borne soils can cause massive dust storms. Where soils are washed in huge amounts into watercourses such as streams and rivers and even oceans, they cause damage to the natural ecosystems. Some of the effects caused by high silt levels in river systems include blocking the river mouth which affects river flow and also renders rivers unsuitable for navigation by boats. Silting also adversely affects fresh- and saltwater marine life such as corals. Coral bleaching, the death of large areas of coral reef, is often associated with soil erosion and run-off due to agriculture. In areas where soil erosion and run-off emanates from agricultural areas, it may also be contaminated by pesticides and fertilizers, which further adversely affects the surrounding riverine and coastal ecosystems. Improved land use measures, such as preservation of natural vegetation along watercourses, contouring of agricultural planting on sloping land and the retention of tree cover, reduces erosion. Trees help in the prevention of erosion because their roots hold soil in place. Many programs have been set up around the world to preserve existing forests or plant new ones.

↑ **Heavy rains from tropical cyclone Gafilo** washed red soils into the river system to the Betsiboka Estuary, Madagascar. The devastating results of this soil erosion can be seen from space. Deforestation practices have removed more than 70 percent of the island's forests in the past century.

← **Improved land management practices** can reduce the impact of logging. To prevent erosion in forests in steep country earmarked for logging, contoured bands of trees have been retained to reduce the impact of run-off.

SAND DUNES

The sand dunes that line many coastal zones are subject to natural erosion from the sea and coastal winds. Where a natural balance exists between the removal of sand and its deposit along beaches by waves, damage to coastlines is minimal. However, many coastal areas are centers of high population and have been damaged by human occupation and exploitation. Reclamation of wetlands and mangroves for roads and housing removes breeding grounds for marine life and reduces habitat. Sand mining is one human activity that has had long-term adverse effects on the vegetation of coastal sand dune communities. In some regions compulsory revegetation programs using introduced species leads to further degradation of natural plant communities since foreign species became weedy, overwhelming indigenous plants.

↑ **Mining has seen the removal of large areas** of foreshore and dune vegetation in coastal areas around the world. Over time and with careful replanting, and the use of physical barriers to control sand erosion, areas may be successfully rehabilitated. The mined dune site north of Richards Bay in Kwazulu-Natal in Africa shows signs of soil stabilization and revegetation.

↟ **Where clearing of natural forests occurs on steep slopes,** erosion can have major and long lasting effects on the landscape. Once the topsoil and plants are removed, regrowth can be slow or even impossible. Here the removal of old growth forests for timber from a high plateau region has left the slopes denuded of both soil and vegetation.

Water conservation and flood protection

Forests and other heavily vegetated land play a vital role in minimizing the rate of run-off and reducing flooding during periods of heavy or sustained rain. Heavily forested regions can act as water reservoirs, retaining huge volumes of water. It has been estimated that 173,000 acres (70,000 hectares) of forest has the water storage capacity of a 35 million cubic foot (1 million cubic meter) reservoir. If forest areas are removed, water flows quickly into the water drainage system rather than being retained in the canopy, roots and leaf litter of the forest. In regions where forests have been removed by logging or through clearing for agriculture, natural drainage systems may fail to cope with the sheer volume of water that flows into them. Floods, often with devastating consequences, can result directly from deforestation in regions around major drainage systems. The damaging effects can be seen along the Yangtze River in China where severe flooding can cause massive destruction of towns and villages. The Yangtze, which is the longest river in Asia, once flowed through heavily forested areas that were cleared to provide timber for fuel and land for agriculture. Much of the deforestation occurred in the 1950s but is continuing today.

MANGROVES AND INUNDATION FROM THE SEA

Floods may also emanate from the ocean, due to tsunamis caused by under-sea earthquakes, or through seasonal high tides or weather events such as hurricanes. Mangroves and coastal swamps play a vital role in mitigating sea floods. Mangroves are found in the intertidal zone between the high and average tide levels. This means that at high tide mangroves are flushed with sea water. At low tide, fresh water flows through the mangroves to the sea. Extensive development of coastlines for industry, tourism and urbanization has seen the removal of vast tracts of mangrove systems and the draining of coastal swamps. There are also concerns that the rising sea levels that may occur through climate change will affect mangroves.

↑ **In many regions that experience heavy or sustained rainfall,** seasonal or frequent flooding is part of the natural weather pattern.

← **The Yangtze River in China** is subject to regular flooding, such as seen here when waters inundated the Jiujiang Economic Development Zone.

↞ **Denuded slopes around the Yangtze River** attest to years of clear-cutting. This practice is the main cause of severe floods that have killed thousands of people along the river in the past decade. Without the forests, more water flows quickly into the river, causing levels to rise more rapidly.

↓ **In a bid to reverse the errors of past development** the people of the island of Palawan in the Philippines are replanting mangroves along their coast. The mangroves will provide habitats for aquatic life and also stem coastal erosion.

Deforestation

Today forests cover approximately 25 percent of Earth's surface, but they once covered more than 50 percent. The decline in forests is directly related to human influence. As the human population has increased and spread around the globe, forests have diminished. The rate of deforestation, literally the removal of forests, has accelerated in the past 100 years. Some forests have suffered more from clearing than others, with dry forests in tropical regions being reduced by more than 70 percent. Around 60 percent of all temperate broadleaf and mixed forests have been removed. Although deforestation is now slowing in many parts of the world, since humans have had a better understanding of the importance and value of forests, clearing still continues at an alarming rate in developing countries in tropical zones. Much of the deforestation that has occurred has been to make room for agriculture, to grow crops or run herds. Logging also continues to lead to vast tracts of deforested landscapes. The loss and degradation of the world's great forests has led to pollution, species and habitat loss and ultimately has contributed to global warming. Importantly, forests lock up about 70 percent of the carbon present in living things. Ongoing removal of forests releases carbon in the form of carbon dioxide into the atmosphere, adding to the greenhouse effect and pressures accelerating global warming.

→ **These sapphire miners** are destroying large tracts of forest inside the Ankarana Reserve in Northern Madagascar. The result is severe degradation of the landscape. The damage from the removal of this forest for mining has broader effects for the region, including removal of topsoil and destruction of habitat and loss of biodiversity. Forest destruction like this is contributing to global climate change.

↓ **Many traditional farmers,** such as these slash and burn tribesmen in a lowland tropical rain forest in the Kikori Basin of Papua New Guinea, regularly employ fire to clear small areas of forested ground to grow crops.

↓ **One of the effects of forest clearing** is the segmentation of large forest areas into smaller regions. Surrounded by industry, agriculture and dwellings, these smaller patches of forest are at risk of invasion by weeds, fire and loss of diversity. In some situations plants can become separated from each other, reducing cross pollination and even removing plants from the pollinators that they depend on for fertilization and reproduction.

↑ **Fire is a naturally occurring phenomenon** that may be caused by lightning strikes or from the lava of volcanoes, such as this one, which is burning through a forest in Hawaii. In regions where fires are a regular natural occurrence, native vegetation has evolved strategies to cope. When fires increase in their frequency through human intervention, tree species may be lost because they are burned too frequently to enable regeneration to occur.

CAUSES OF DEFORESTATION

Deforestation may be caused by human intervention, in the form of clearing forested areas for urban development or agriculture, or it may be natural, such as wildfires, volcanoes, hurricanes or tornadoes. Fire is a major factor affecting forests. In many vegetative communities fire is a regular occurrence. Changes in climate, particularly higher temperatures and long, dry spells, along with the growth of urban development, are placing native forests around the world under more regular threat of fire. If the rate of burning is too frequent, plant communities are devastated rather than invigorated by fire. Trees in particular may need many years between fire events to ensure their survival either through regeneration or through reproduction and seeding. If young plants are repeatedly burned before they can mature, flower and seed, the species will soon die out leaving a degraded landscape.

Reforestation

Scientists report that carbon dioxide levels in the atmosphere are higher today than at any time in the past 650,000 years, and that these increased levels are causing climate change. Trees lock up carbon and the protection of forests from unchecked logging and wholesale clearing is a vital part of counteracting the increase in carbon dioxide emissions. As well as bringing pressure to bear to stop clearing, particularly of tropical rain forests, conservationists want more trees to be replanted around the globe. The timber industry is one entity capable of large scale reforestation. Managed forests, where cleared areas are replanted for a future crop, show that a sustainable approach to timber supply is possible. Many other successful reforestation projects result from the efforts of community or school groups replanting small degraded areas. These programs are championed through set planting days, such as Arbor Day when large public tree planting projects are carried out by groups of volunteers. Through ongoing tree planting, it is hoped that environments can be saved from further damage, and that endangered plant and animal species will be given back vital habitat lost through land clearing and land degradation.

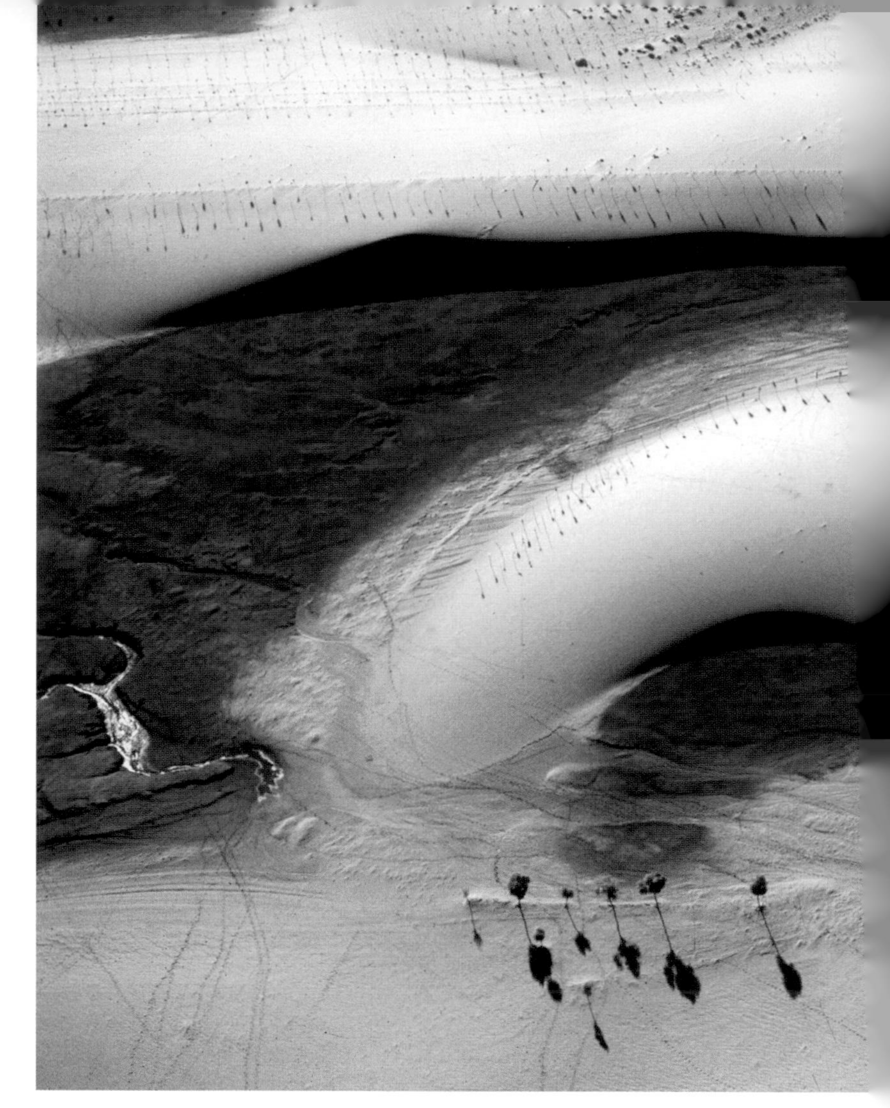

CARBON TRADING

One way to encourage reforestation is to place a monetary value on trees as they grow, rather than only after harvest. This has been made possible through an appreciation of the tree's ability to lock up carbon and so help reduce the amount of carbon dioxide in the atmosphere. A system known as carbon trading—fast becoming part of the global economy—translates the environmental value of trees into a tradable commodity. In order to offset carbon emissions, polluters pay for their emissions with carbon credits obtained by planting trees. In most cases, the tree planting occurs in regions away from the area of pollution. Credits are only given for tree planting that occurs over a large area and where the trees that are planted live and grow.

SUCCESSION IN A CONIFEROUS FOREST

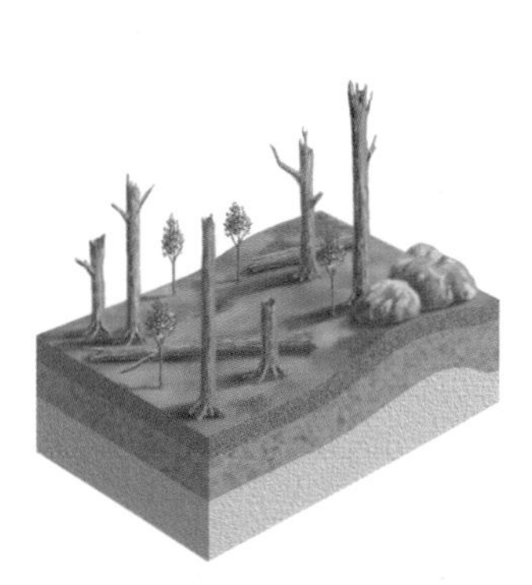

Birch saplings emerge from the burned soil

The birch trees grow to full size

After around 60 years pines take over from the birch

After around 200 years spruces take over from the pines

↑ **China's Great Green Wall** is a replanting initiative of the Chinese government that began in 2001. The project's aim is to plant a 2800-mile (4480-km) shelterbelt over 70 years to hold back the moving sands of the Gobi Desert.

← **Following a forest fire,** trees may be completely razed to the ground. However, over time the forest will grow back in a process known as succession.

↑ **Trees must be nurtured** to ensure survival after replanting. This one is surrounded by a plastic sleeve to stop animals damaging the bark or roots.

↖ **Reforestation at work**. Here a crew for the Costa Rican environmental group Arbofilia, plants trees in the degraded Puriscal region.

← **Sustainable forests** such as this plantation of Sitka Spruce in Scotland, are managed so that cleared areas are reforested then grown to maturity, providing a future timber crop.

Factfile

Factfile

Tree shapes 284

Leaf shapes and arrangement 285

Flowers and infloresences 286

Flower shapes 286

Fruit types 286

Major forest biomes of the world 287

Important families and genera of trees 288

Critically endangered trees 293

Tree records 293

TREE SHAPES

The greatly varied outlines and architectures of trees are a reflection of the genes that guide their sequence of branching; dominance or otherwise of the leading shoot; shape, size and spacing of leaves; and even the way a shoot behaves following production of an inflorescence.

LEAF SHAPES AND ARRANGEMENT

STRUCTURE AND OUTLINE

When identifying a tree, one of the obvious features is the overall outline of the leaf. Botanists have devised a terminology for a wide range of shapes, the English terms closely following the traditional Latin. Here are some of the basic shapes.

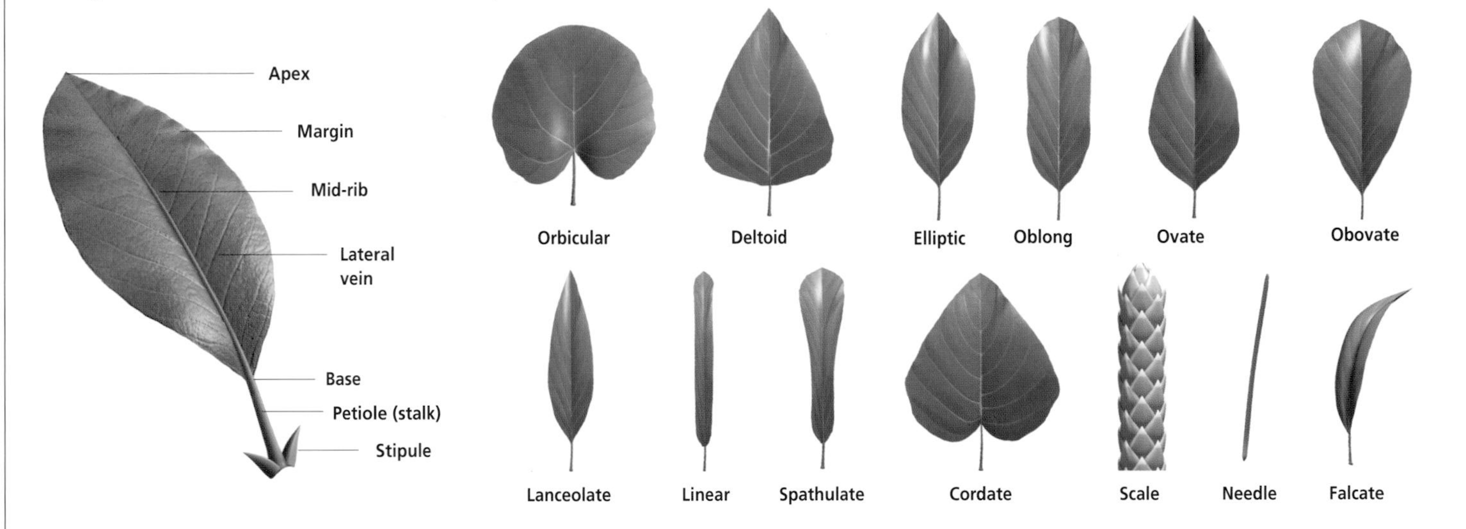

MARGIN

Leaf margins are features that need to be considered when identifying trees. For example, some families of trees commonly have toothed leaves, while others never have them.

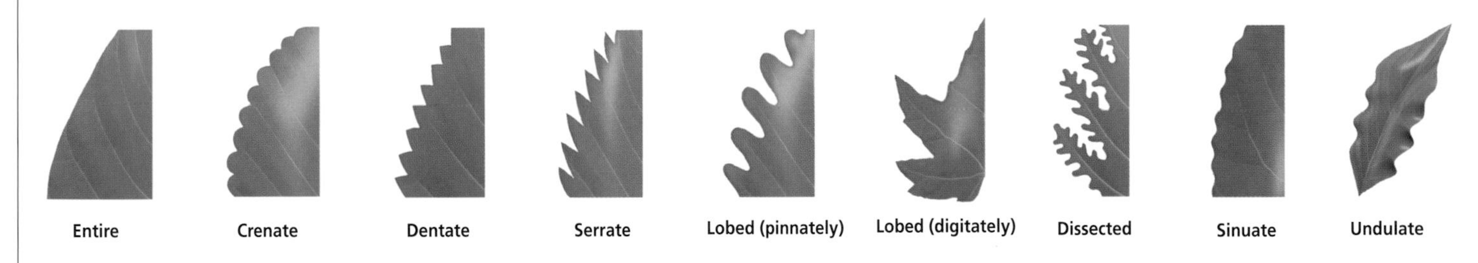

ARRANGEMENT

The way leaves are arranged on a stem is characteristic of a species, often of its genus as well, or even a whole plant family. The most basic distinction is between an alternate or spiral arrangement and opposite pairs (or whorls of three or more).

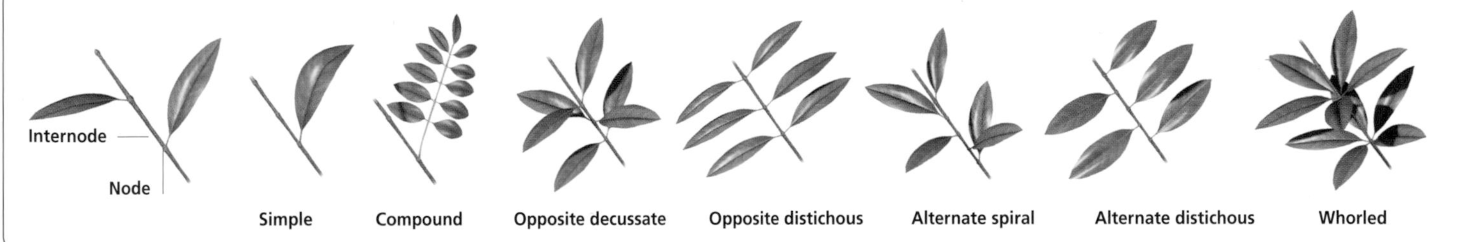

FLOWERS AND INFLORESCENCES

The structures and shapes of flowers are almost infinitely varied and these diagrams show only some basic features. The first shows a simple flower type with the four basic whorls of organs: sepals, petals, stamens and pistils. Next are shown some of the major types of inflorescence (specialized flower-bearing branch).

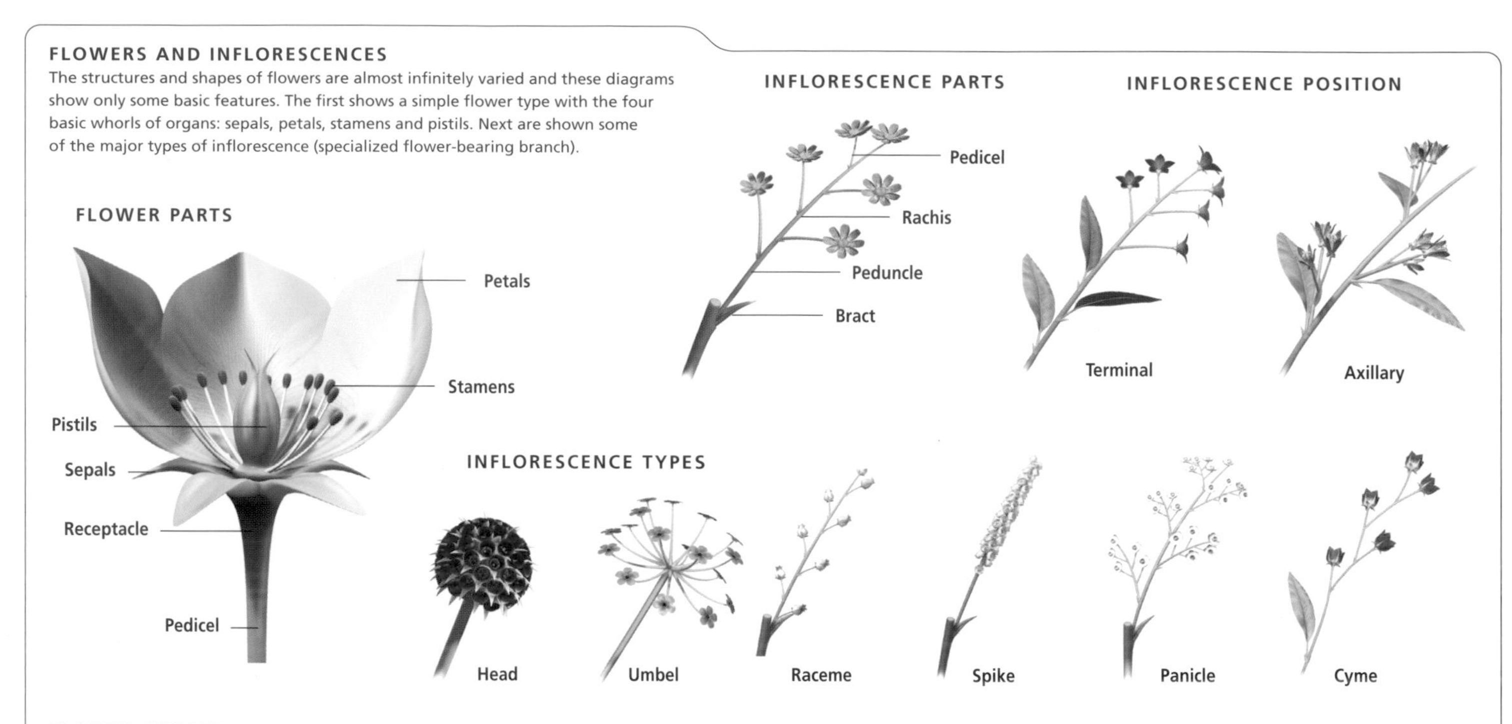

FLOWER SHAPES

These are only some of the large range of shapes that flowers can have. In many flowers the petals are fused by their edges to form a tube, bell, urn or trumpet shape. Shapes and colors are adaptations to attract particular pollinators.

FRUIT TYPES

The fruits, the seed-containing organs of flowering plants, also come in an almost endless range of shapes and sizes, of which these illustrations show some of the most basic. Fruits are classified according to many criteria, including whether they are dry or fleshy.

MAJOR FOREST BIOMES OF THE WORLD

Plant ecologists classify vegetation by criteria such as structure, which encompasses tree density, canopy height, presence of lower layers such as understory trees, shrubs and grasses, and presence of growth forms such as lianes, epiphytes or parasitic plants. Biomes are at the highest level of classification. They are the geographically largest biological communities, comprising a characteristic type of vegetation, as seen here.

- Boreal forest
- Temperate coniferous forest
- Temperate broadleaf deciduous forest
- Temperate southern hemisphere broadleaf forest
- Temperate woodlands
- Tropical rain forest
- Tropical dry deciduous forest
- Savanna
- Mangroves

Boreal forest

Temperate coniferous forest

Temperate broadleaf deciduous forest

Temperate southern hemisphere broadleaf forest

Temperate woodlands

Tropical rain forest

Tropical dry deciduous forest

Savanna

Mangroves

IMPORTANT FAMILIES AND GENERA OF TREES

These include some of the largest tree genera (in terms of number of species) as well as some smaller genera that may be important elements of some widespread vegetation types, or are significant as cultivated trees

Genus	Number of species	Geographical range	Characteristic features	Significance
CONIFERS				
Araucariaceae *Ancient conifer family, most abundant in age of dinosaurs, now has very fragmented occurrence, mainly southern hemisphere*				
AGATHIS Kauris	21	Western Pacific region from Sumatra to Philippines, Fiji, New Zealand and northeastern Australia	Mostly massive trees with smooth or flaky bark; distinctive flat, leathery leaves; ball-like seed cones with smooth outer surface	Much exploited for timber in the past, but few mature trees now remaining outside reserves; resin from trunks has various uses
ARAUCARIA Monkey Puzzle, Norfolk Island Pine	19	Fragmented distribution: 13 species in New Caledonia, 1 in Norfolk Island, 2 in eastern Australia, 2 in New Guinea (1 shared with Australia), 2 in South America	Strikingly symmetrical trees with whorled branches; spine-tipped leaves; pollen cones large and elongated; seed cones large with prickly scales	Widely planted for ornament including indoor use
Cupressaceae *Cypress family: most diverse and widely distributed conifer family, includes former family Taxodiaceae; some deciduous*				
CALLITRIS Australian cypresses	19	17 species in Australia (widespread), 2 in New Caledonia;	Some features show parallels with northern hemisphere *Cupressus*, but leaves and cone scales in whorls of 3	Timber trees in some inland regions of eastern Australia; whitish resin (sandarac) formerly used in pharmaceuticals
CUPRESSUS True cypresses	c.28	Scattered around warmer temperate regions of the northern hemisphere, principally China-Himalaya and western North America	Fine, dense foliage with tightly packed opposite scale-leaves, 4-ranked on tiny branchlets; woody globular seed cones with few scales	Widely grown for ornament and shelter from wind; some species have cultural associations with graveyards
JUNIPERUS Junipers	c.50	Northern hemisphere from subarctic regions to some tropical mountain ranges	Evergreen trees; foliage like *Cupressus* or leaves small prickly needles; cones berry-like with fleshy scales fused together	Abundant in poorer woodland and scrub in some regions, e.g., central plateaus of North America; "berries" of *J. communis* flavor gin
Pinaceae *Pine family: needle-bearers, all from northern hemisphere; evergreen except for the larches (*Larix*) and* Pseudolarix				
ABIES Firs	c.50	Northern hemisphere (in Africa confined to mountains of northern Algeria), south to eastern Himalaya region and high mountains of Guatemala	Tall, stately conifers with spirelike habit and short, flattened needles crowded on branchlets; seed cones upright, shattering at maturity	Dominant scenery element in some cold mountain ranges, e.g., Cascades and Sierra Nevada of western USA; some timber trees; also popular as ornamentals
PICEA Spruces	33	Northern hemisphere from above arctic circle (abundant in taiga) to Tropic of Cancer (mountains of Taiwan and Mexico)	Similar in habit to *Abies* but needles mostly not flattened, attached to peglike outgrowths of twigs; seed cones pendent, not shattering	Most important lumber- and pulp-yielding trees of far northern regions; popular as ornamentals
PINUS Pines	111	Northern hemisphere, from arctic circle to tropics excepting Africa (apart from far northwestern Africa); just crossing the equator in Sumatra; most diverse in Mexico	Evergreen trees with unique shoot structure with 3 kinds of leaves, mostly long needles grouped into bundles of 2–6 (rarely 1); seed cones with woody or leathery scales, developing over 2–3 years	Major source of softwood lumber and pulpwood, from cool-temperate native forests, or from plantations in southern hemisphere countries; resin of some species the original "turpentine;" widely planted for shelter and ornament
Podocarpaceae *Podocarp family: most diverse in southern hemisphere and tropics; fleshy fruitlike cones*				
PODOCARPUS Podocarps	94	Mainly tropical forests, north to Japan and eastern Himalaya, south to southern parts of all major southern hemisphere landmasses	Shrubs to tall trees with flattened, thick leaves; seed cones reduced to fleshy stalks with one or few large berry-like seeds	Some grown for ornament; botanically interesting as relics of ancient Gondwanan forests

IMPORTANT FAMILIES AND GENERA OF TREES *Continued*

Genus	Number of species	Geographical range	Characteristic features	Significance
FLOWERING PLANTS: DICOTYLEDONS				
Aceraceae *Maple family: single genus,* Acer, *of northern hemisphere; closely allied to the larger family Sapindaceae (arguably not distinct)*				
ACER Maples	c.120	Northern hemisphere, majority in east Asia, 14 in Europe; only 12 in North America but these include important trees	Deciduous, opposite leaves, usually toothed or lobed; flowers small, clustered; fruit dry, consisting of 2 joined nuts (samara) each terminating in a "wing," wind-carried	Widely planted for ornament, famed for fall color; few species valued for timber apart from *A. saccharum*, which also yields maple syrup
Aquifoliaceae *Holly family: all but a few species belong to the cosmopolitan genus* Ilex, *mainly evergreens of tropics and subtropics*				
ILEX Hollies	c.400	Worldwide except Australia (only 1 species in far north) and boreal regions; China accounts for over half the total species	Mostly small trees, evergreen or deciduous; leaves often leathery, sometimes with spine-tipped teeth; flowers small, greenish; fruit a small berry	Many grown for ornament; some have cultural associations such as European Holly (*Ilex aquifolium*), its red berries displayed at Christmas
Betulaceae *Birch family: few genera of mostly deciduous trees, catkin-bearing and wind-pollinated, important in cool-temperate regions*				
ALNUS Alders	35	Predominantly northern hemisphere (Africa only in northwest), from arctic circle south to eastern Himalaya, in Americas south through Andes to Argentina	Deciduous, leaves alternate, doubly toothed; male and female flowers in separate branched catkins; wind-borne pollen and seeds	Pioneer trees on river bars, glacial moraines, volcanic ash etc.
BETULA Birches	c.50	Northern hemisphere from above arctic circle to Thailand in Eurasia (Africa only in high mountains of northwest) and southern USA in Americas	Bark distinctive, peeling in papery flakes or sheets; deciduous, leaves like *Alnus* but mostly thinner, shorter and proportionately wider; male and female flowers in separate catkins, in leaf axils; wind-borne pollen, female catkins disintegrate into small papery seeds	Major broadleaf trees of boreal forest, rapidly colonizing by seed; papery bark has many uses; popular ornamental trees
Bignoniaceae *Trumpet-flower family: mainly tropical climbers and trees, many genera and species with showy trumpet-shaped flowers*				
TABEBUIA Trumpet-trees	c.100	Central and South America (south to Argentina), West Indies	Mostly deciduous (dry-season) trees of rain forest to savanna; leaves digitately compound, opposite; flowers large, colorful, trumpet-shaped, in terminal clusters	Some important timber trees in South America (wood very durable); some grown as ornamental flowering trees throughout tropics
Cannabaceae *Hemp family, includes hemp and hops but recent evidence leads to adding some tree genera, formerly placed in Ulmaceae*				
CELTIS Hackberries, nettle-trees	c.100	Tropics to warmer temperate regions of world, north to southern Europe, Caucasus, Japan, in North America to New York State	Evergreen or deciduous; simple, alternate leaves on zigzagging twigs; small petaless flowers; fruit a small berry-like drupe, often ripening as blackish	The deciduous temperate species often planted as shade trees; close relationship to *Cannabis* a recent discovery
Casuarinaceae *She-oak family: only 4 genera of trees with conifer-like branches but closely allied to Betulaceae; majority Australian*				
CASUARINA She-oaks	17	Australia and western Pacific region, north to Southeast Asia	Trees with highly modified leaves, in whorls of 5–20 fused to faces of twigs, producing needle-like foliage; male and female flowers in catkins on separate trees, wind-pollinated; seeds small, wind-carried	Another 60 species in the very similar endemic Australian genus *Allocasuarina* (also called she-oaks); planted for timber, shelter and firewood
Ebenaceae *Ebony family, mainly tropical, only 1 small genus in addition to the very large* Diospyros				
DIOSPYROS Ebonies, persimmons	c.475	World tropics and warmer temperate regions, north to Mediterranean, Japan and eastern USA; majority in Asia, Madagascar and Africa	Mostly evergreen, crooked-limbed trees and shrubs with hard, strong wood streaked with black pigment; simple alternate leaves; flowers greenish, in leaf-axils; fruit a berry, sometime large	Wood valued for strength, density and coloring, used for carving and golf clubs; fruit mostly edible, some species of commercial significance
Fabaceae *The legumes, third-largest flowering plant family, c.18,000 species, most abundant in temperate zones and drier tropics*				
ACACIA Wattles acacias, mimosas	c.1500	Warmer regions of the world, but majority (950 species) Australian, others mainly African and American	Trees or shrubs; true leaves bipinnate but commonly replaced by phyllodes resembling simple leaves; flowers tiny, in dense globular heads or short spikes; fruit a typical leguminous pod	Often abundant in arid regions, nitrogen-fixing by bacteria in root nodules; some quick-growing plantation trees for wood and tanbark; some species yield gum arabic

IMPORTANT FAMILIES AND GENERA OF TREES *Continued*

Genus	Number of species	Geographical range	Characteristic features	Significance
DALBERGIA Rosewoods, Palisander	c.100	Tropics, most in Americas and Africa	Evergreen trees or lianes; alternate pinnate leaves with few leaflets; clustered pea-flowers; pods flat, not splitting to release seeds	Timber highly valued, durable and richly grained and colored, used for carving, cabinetwork, knife handles etc.
ERYTHRINA Coral trees, coral-beans	112	Tropics and subtropics of world, majority in the Americas	Deciduous; prickles on branches and leaf stalks, leaves compound with 3 leaflets, like the closely related beans (*Phaseolus*); flowers showy, bird-pollinated, in dense terminal spikes	Some grown for ornament; red seeds used for handcrafts and souvenirs
Fagaceae *Beech family, also oaks and allied genera: mainly northern hemisphere, the dominant broadleaf trees in some regions*				
CASTANOPSIS	c.120	Tropical and subtropical Asia and Malay Archipelago, north to Japan and central China	Evergreens, closely allied to the deciduous chestnuts (*Castanea*) with similar catkins of male and female flowers; nuts enclosed in spiny or bristly cupules	Important components of east Asian hill forests, yielding timber and edible nuts
FAGUS Beeches	10	Temperate northern hemisphere, south to mountains of northern Vietnam and northeast Mexico, north to southern Scandinavia and southeastern Canada	Deciduous with prominent scaly winter buds, thin leaves; separate globular male and female catkins on fine stalks; small angled nuts enclosed in shaggy cupules	Locally abundant trees of deciduous broadleaf forest; timber widely used; nuts (beech-mast) food for animals
NOTHOFAGUS Southern beeches	34	Southern hemisphere, confined to Australasian region (New Zealand, southeastern Australia, New Caledonia, New Guinea) and temperate South America (southern Chile and Argentina)	Evergreen or deciduous; leaves like *Fagus* in some species, leathery or smaller in many others; male and female catkins both short; small angled nuts among shaggy bracts	Dominant trees of temperate rain forest in Chile, New Zealand, Tasmania: exploited for timber especially in Chile
QUERCUS Oaks	c.530	Mainly northern hemisphere, southern Scandinavia to northwestern Africa and through tropical Asia to Indonesia; and from eastern Canada to mountains of Colombia; most diverse in China and Mexico	Evergreen or deciduous; leaves simple, alternate, usually toothed or lobed; male catkins delicate, drooping, female thick; fruit a typical acorn partially enclosed in a cupule	Many uses: timber famed for strength and durability; tannin-rich bark and twig galls used for tanning; acorns for animal feed; often planted for shade and ornament
Lauraceae *Laurel family: ancient (pre-dicot) family of mostly evergreen rain forest trees, spicy aromatic oils in bark, leaves and fruit*				
LITSEA	c.400	World tropics and subtropics, most in Southeast Asia and Malay Archipelago, south to southeastern Australia; few in Africa or Americas (1 north to southeastern USA)	Evergreen trees; leaves simple, alternate, leathery; flowers in profuse small heads, mixed male and female; fruit a drupe in a fleshy cupule	Common trees of tropical rain forests and other humid sites; yield timber, essential oils, some minor fruits
PERSEA	c.200	Tropical and temperate America, north to Virginia, south to southern Chile; eastern and southern Asia, north to central China	Evergreen trees; leaves simple, alternate, leathery; flowers small in delicate stalked panicles; fruit 1-seeded, with oily flesh	Includes the Avocado (*P. americana*) and some other species grown for fruit; timber of many species exploited
Magnoliaceae *Magnolia Family: ancient (pre-dicot) family, mostly tropical evergreens, some temperate deciduous; showy large flowers*				
MAGNOLIA Magnolias	c.220	Tropical and eastern Asia and Malay Archipelago (south to New Guinea); eastern North America and Tropical America south to southern Brazil; many tropical ones formerly classified as *Michelia*, *Talauma*, *Manglietia*	Small to large trees, most evergreen, few deciduous; leaves simple and entire with stipules encircling twigs and enclosing terminal bud; flowers often large, multi-petaled, with many separate carpels	Some tropical species yield timber; some popular as ornamentals, both evergreen and deciduous (many hybrids)
Malvaceae *Mallow and Hibiscus family, now widened in scope to include kapoks, baobabs, dipterocarps, sterculias, basswoods etc.*				
BOMBAX Silk-cotton trees	20	Tropical Africa, Asia and Australia; tropical American species are separated as genus *Pseudobombax*, also with c.20 species	Large trees, fully deciduous in dry season when flowering; trunks prickly; flowers large, showy, with 5 fleshy petals and "shaving-brush" group of many stamens; fruit a large capsule, seeds embedded in mass of kapok	Striking emergent trees of rain forests and vine thickets; planted for ornament; trunks for dugout canoes; soft timber for packing-cases etc.; seed fiber for stuffing; medicinal uses

IMPORTANT FAMILIES AND GENERA OF TREES *Continued*

Genus	Number of species	Geographical range	Characteristic features	Significance
DIPTEROCARPUS Keruing, Yaang	c.70	Tropical Asia, Malay Archipelago excluding New Guinea	Large trees, evergreen or dry season deciduous; leaves simple, often large; flowers hibiscus-like but with calyx 2-lobed, lobes developing into large wings like propeller blades on nutlike fruit	Characteristic dipterocarps, often dominant trees in rain forest or tropical deciduous forest; yield valuable strong timber, heavily exploited
SHOREA Sal, Meranti, Balau	c.360	Tropical Asia (northern India to southern China); Malay Archipelago excluding New Guinea (majority of species)	Like *Dipterocarpus* but calyx and nut with 3 large and 2 small wings	Dominant and often largest trees of rain forests; most important source of rain forest timbers in Asia
STERCULIA Tropical chestnuts	c.150	World tropics and subtropics	Mostly deciduous; leaves simple (often palmately lobed) or palmately compound; flowers star-shaped to bell-shaped; fruit of 5 separate carpels, splitting at early stage of development to reveal seeds and colored inner surface	Some lightweight timbers; bark fiber used for rope, nets etc.; seeds, roots, leaves variously edible or medicinal; seed oil used for cooking or lighting; bark gums used variously
Moraceae *Mulberry and Fig family, mostly tropical trees, shrubs and climbers; milky sap; flowers small, crowded, on or inside fleshy spikes*				
FICUS Figs	c.1000	World tropics and subtropics, throughout continents and most islands	Deciduous and evergreen, including many with "strangler" growth habit; leaves simple, entire or lobed, stipule enclosing terminal bud, sap milky; flowers and fruits minute, enclosed in a unique "inside-out" inflorescence (fig) requiring tiny wasps for pollination	Some figs food for humans, many others important to mammals, birds, reptiles, even fish; shoots and leaves of some edible or medicinal; widely planted for shade and ornament including indoor use
Myrtaceae *Myrtle family, diverse in American and Asian tropics also temperate Australia and South America, much less diverse in Africa*				
EUCALYPTUS Eucalypts, gum trees	c.800	Australia (throughout) except for 9 species in New Guinea and southeastern Indonesia, and 1 endemic to southern Philippines and northern New Guinea	Small to very tall trees, all but a few evergreen; leaves mostly narrow, leathery, smooth, containing essential oils; flowers with petals and/or sepals fused into a cap which falls to reveal massed stamens, showy in many species; fruit a woody capsule	Dominant trees of most Australian forests, often thriving in very poor soils, important for dense, strong timber; planted in many warmer countries for timber, firewood, oil, shelter, ornament
EUGENIA Stoppers	c.550	Mainly tropical and South America (north to Florida); few in Africa, Asia, Australia	Shrubs to large trees, evergreen or deciduous; opposite smooth-edged leaves; small usually white flowers; single-seeded fruit with succulent, aromatic flesh	Many with edible fruit, some cultivated for fruit; some exploited for hard, heavy timber
MELALEUCA Paperbarks, honey-myrtles	220	Australia (almost throughout); small number of species extending to New Caledonia, Malay Archipelago and Southeast Asia	Evergreen shrubs and trees, some with papery, many-layered bark; leaves thick-textured, often aromatic; flowers densely clustered, often colorful, with massed stamens, these fused basally into 5 bundles; fruit small woody capsules	Many grown as ornamentals or for shade or shelter; essential oils (e.g., "tea-tree oil") obtained from some; *M. quinquenervia* has become major weed of Florida's Everglades
SYZYGIUM Jambu, Water-apple, Clove	c.1000	Wet tropics and subtropics of Asia and Australasia; few species in Africa	Evergreen trees; leaves opposite, smooth; flowers with many stamens; fruit fleshy, crisp, single-seeded, with persistent sepals at apex	Many with edible fruit, some cultivated for fruit; dried flower buds of clove (*S. aromaticum*) a major spice; some timbers, ornamentals
Oleaceae *Olive family, worldwide, most diverse in tropics, includes some popular flowering shrubs of temperate gardens as well as the Olive*				
FRAXINUS Ashes	c.60	Mainly temperate Eurasia, North America and north Africa; few in tropics, south to highlands of Indonesia and Guatemala	Evergreen or deciduous; leaves opposite and pinnate; flowers small but profuse, petals 4 or absent; fruit a winged, narrow nut ("key")	Some important timber trees: wood strong and resilient, favored for tool handles, oars, etc.; much planted as shade trees
OLEA Olive and relatives	c.40	Warm-temperate to tropical Africa, Asia, Australia, Pacific Islands	Evergreen trees, leaves opposite, simple, often thick-textured; flowers small, 4-petaled; fruit a drupe with oily flesh and very hard stone	Includes the cultivated Olive (*O. europaea*) probably derived from wild north African species—major source of edible oil; other species yield very heavy, strong timbers
Rosaceae *Rose family, large and diverse mainly in northern hemisphere; many popular flowers and fruits of temperate climates*				
PRUNUS Plums, cherries, peach, apricot etc.	c.350	Tropical and temperate regions of the world; few in South America, only 1 in Australia (in far northeast)	Deciduous or evergreen; leaves alternate, simple, usually toothed and bearing nectar glands (mostly on leaf stalks); flowers often colored and showy, with 5 separate petals, many stamens, only 1 carpel; fruit a drupe with usually juicy edible flesh and hard stone	Edible fruits include the plums, cherries, peach, nectarine, almond, apricot; numerous ornamental species and cultivars, grown for floral display and amenity; some used for timber

IMPORTANT FAMILIES AND GENERA OF TREES *Continued*

Genus	Number of species	Geographical range	Characteristic features	Significance
SORBUS Mountain ashes	c.250	Temperate Eurasia and North America, from subarctic regions to highlands of Mexico and Thailand	Deciduous; leaves alternate, pinnate or simple, toothed or lobed; flowers in large panicles, small, 5-petaled, cream; fruit like miniature apples or pears (pomes)	Closely allied to apples and pears; a few species valued for fruit but most grown for ornament; much folklore associated with European *S. aucuparia*
Salicaceae *Willow family , traditionally only willows and poplars but enlarged on recent evidence to include many tropical tree genera*				
POPULUS Poplars	c.90	Mainly temperate Eurasia, North America, north Africa, extending into tropical highlands in east Africa	Deciduous trees with resinous winter buds; leaves alternate, simple, toothed, often broad in proportion to length; flowers small, in male and female catkins on separate trees; seeds like thistledown, from small capsules	Lightweight timber used for matches, plywood, crates, clogs; usually plantation-grown; widely planted for shade and amenity or for erosion control
SALIX Willows	c.500	Mainly temperate, a few species extending to South Africa, South America, tropical Asia; prominent in Arctic and subarctic regions	Deciduous (few exceptions) trees and many shrub species; leaves alternate, simple, often narrow and finely toothed; flowers, fruits and seeds as in *Populus* but smaller	Colonizers of riverbanks and sandbars; light but close-grained wood has local uses, e.g. cricket bats, clogs, clothespegs; young canes used for basketry, lobster traps, etc.; widely planted
Ulmaceae *Elm family: smallish family of trees, northern nemisphere; mostly deciduous, the leaves toothed; fruit a small flat nut*				
ULMUS Elms	c.40	Northern hemisphere, mainly temperate, south to Thailand and Costa Rica; absent from Africa except far northwest	Deciduous or semi-deciduous; leaves alternate, simple, toothed, asymmetric at base; flowers small; fruits small, flattened, papery (carried by wind)	Once major timber trees and scenic elements of Europe and eastern USA but devastated by Dutch elm disease in mid-20th century
FLOWERING PLANTS: MONOCOTYLEDONS				
Agavaceae *Agave family: scattered around warmer parts of world but larger-growing plants nearly all from the Americas, especially Mexico*				
YUCCA Yuccas	c.40	North and Central America, from northern USA to Guatemala	Evergreen swordleafed small trees to herbaceous perennials mostly with leaves spine-pointed; flowers white, bell-shaped, in large, mostly upright panicles	Fiber from leaves has many local uses; all planted for ornament or as "living fences"
Arecaceae – *Palm family: large family of trees, shrubs and climbers with large fronds, found throughout tropics and subtropics, very diverse*				
ATTALEA Neotropical oil palms	29	Central and South America and West Indies, from Mexico to Argentina	Medium-size to large palms with massive pinnate fronds and flowering panicles; 1–3-seeded fruits densely packed on spikes	Dominant large palms in many parts of the Americas, allied to the Coconut (*Cocos*); major source of seed oil in some regions; fronds used for thatching
PHOENIX Date palms	17	Africa, Canary Islands, southern Asia; one species in Europe (Crete)	Small to large palms with pinnate fronds, lower leaflets modified into sharp spines; male and female flowers on separate trees; fruit a 1-seeded drupe with hard stone	Dates, from *P. dactylifera*, a major fruit, traded worldwide in dried state; some species widely planted as ornamentals
Asphodelaceae *Asphodel family: mostly herbaceous plants from Africa, Mediterranean and western Asia; some succulents especially Aloe*				
ALOE Aloes	c.360	Africa, Madagascar, Arabia, Canary Islands	Evergreen swordleafed plants, most shrubby or herbaceous, few becoming trees; leaves succulent, usually with prickles on edges; flowers tubular, showy, in dense spikes, bird-pollinated	Viscous sap from leaves of some used medicinally and in cosmetics; planted for ornament and collected by succulent plant enthusiasts
Pandanaceae *Pandanus family: only 3 genera of which* Pandanus *largest; tropics, majority in Southeast Asia; absent from Americas*				
PANDANUS	c.700	Tropics of Asia, Africa, Australia and Pacific; the great majority in Malay–Philippines region and Madagascar	Evergreen swordleafed trees, shrubs and climbers; leaves with prickly margins and ribs, in prominent spirals on stems; male and female flowers on separate plants; female developing large, complex fruit heads	Prominent in vegetation of swamps and sea coasts; many local uses including leaves for thatching, basketry, food flavoring; fruit of some used for food

CRITICALLY ENDANGERED TREES

Below are listed just 10 examples of the world's many tree species in danger of extinction. They do not include trees that have been successfully and widely grown, such as the Flamboyante (*Delonix regia*), Franklin Tree (*Franklinia alatamaha*), Ginkgo (*Ginkgo biloba*), Dawn Redwood (*Metasequoia glyptostroboides*), and even the recently discovered Wollemi Pine (*Wollemia nobilis*).

Critically endangered trees include some that botanists and land managers have kept under surveillance for decades; but there are many others, especially in tropical rain forests, that have only recently been discovered and identified. Some have proved impossible to find again following their naming as new species, because the forests they grew in have since been cleared—and it is certain that many more have become extinct in recent years without ever becoming known to science.

Some known "hot-spots" of tree endangerment include: oceanic islands such as the Canary Islands, St. Helena, Indian Ocean islands—especially Madagascar and Mauritius—and the Hawaiian Islands; mountains of Vietnam and adjacent region of southern China; plateaus of Mexico and Central America; and coastal forest of southeastern Brazil. Rain forests of the Amazon and central Africa also undoubtedly contain many trees that are under threat from extinction, but their vastness and (in Africa) political instability makes botanical monitoring almost impossible.

Abies beshanzuensis **Baishanzu Fir**
A handsome tall fir from Mt. Baishanzu in China's southeastern Zhejiang Province, this species is known in the wild only as three adult trees, although it has been taken into cultivation. It was named as recently as 1976.

Cercocarpus traskiae **Catalina Island Mountain Mahogany**
Surviving only as one small copse in a dry, rocky gully on south California's Santa Catalina Island. A long-lived evergreen of dense, shrubby habit, it can reach 25 feet (8 m) in height. The unique vegetation of this island has suffered great damage by introduced animals, but lately there has been much work put into its restoration.

Cupressus dupreziana **Tarout Cypress**
Isolated on one rocky plateau of the Tassili N'Ajjer massif in Saharan Algeria; there are only about 150 live trees surviving; very few seedlings or saplings have ever been found. Its population has long been reduced by firewood gathering, sheep and goat browsing, and lowering of the water table.

Dimorphandra wilsonii
Known only from 10 mature trees and a few saplings in the wild in its native Minas Gerais State in southern Brazil, growing in the *cerrado* (a type of savanna). A handsome leguminous tree related to the honey-locusts (*Gleditsia*), it is threatened by clearing for pasture and by charcoal-burning.

Elaeocarpus bojeri **Bois Dentelle**
Endemic to Mauritius, fewer than 10 small trees survive in a small patch of cloud forest; threatened mainly by invasion of exotic plants, especially the Cherry Guava (*Psidium cattleianum*) which competes with native vegetation on many other oceanic islands.

Idiospermum australiense **Ribbonwood**
Surviving only in a few small pockets of coastal lowland rain forest in the Daintree region of northeast Queensland, Australia, its habitat greatly reduced by clearing for sugarcane-growing, grazing and housing. Remarkable relict tree in the pre-dicot family Calycanthaceae; the only member of the family in Australia and only distantly related to its other members.

Kokia drynarioides **Hawaii Cotton Tree**
From dry lava forests on the leeward side of the "Big Island" (Hawaii), this species survives as fewer than 10 small trees. In fact, of the four known *Kokia* species, all Hawaiian, two are rare in the wild, one only survives precariously in cultivation, and one is totally extinct. They are attractive trees or shrubs with colorful Hibiscus-like flowers.

Pleodendron costaricense
Discovered only in 1998 and identified and named in 2005, this rain forest tree of up to 120 feet (35 m) tall is known to exist as only two mature trees and one sapling beside a road in the lowlands of Costa Rica's Puntarenas Province. The only other species of *Pleodendron* is likewise an endangered tree from Puerto Rico, some 1200 miles (1900 km) away. The trees belong to the primitive pre-dicot family Canellaceae.

Trochetiopsis erythroxylon **St. Helena Redwood**
A small tree once common on higher parts of St. Helena, but so heavily exploited for timber by early European settlers that it came close to extinction by the early 18th century. Now extinct in the wild, a few trees survived in island gardens and seedlings have been raised for replanting in the wild.

Zelkova sicula **Sicilian Zelkova**
One of Europe's few endangered trees, this recently discovered species survives as only about 200 trees scattered along a stream bank in the remote Iblei Mountains of Sicily. It is the westernmost species of the genus *Zelkova*, which has a fragmented distribution in Japan, China, the Caucasus, Crete and Sicily.

TREE RECORDS

Tallest tree Nicknamed "Hyperion,' a Coast Redwood (*Sequoia sempervirens*) in Redwood National Park in the far northwestern corner of California, which was measured in 2006 at 379.1 feet (115.55 m).

Largest diameter trunk Often cited is El Árbol del Tule in Oaxaca, Mexico, a massive specimen of Montezuma Cypress (*Taxodium mucronatum*) with a girth of 118.8 feet (36.2 m) at its base—equating to a diameter of 37.8 feet (11.5 m). But it has many protruding buttresses that inflate this measurement, and the stoutest Giant Sequoia at 29.5 feet (9 m) diameter has almost the same cross-sectional area.

Largest timber volume This record goes to a Giant Sequoia (*Sequoiadendron giganteum*), specifically the famous General Sherman tree in California's Giant Sequoia National Park. At 274 feet (83.6 m) tall and 27.1 feet (8.25 m) in diameter, it has a trunk volume calculated at 54,380 cubic feet (1540 cubic m). But with bark up to 18 inches (45 cm) thick, the timber volume is only around 80 percent of the trunk volume, or approximately 43,000 cubic feet (1200 cubic m)—enough to build over 150 woodframe houses.

Widest canopy The largest continuous tree canopy is believed to be that of a many-trunked Banyan (*Ficus benghalensis*), though recorded measurements are mostly vague and outdated. One in northern India is reported to cover an area of 5.2 acres (2.1 hectares) which, if it were approximately circular, would equate to a diameter of 540 feet (165 m). As for a canopy supported by a single trunk, it would be almost impossible to discover what the world's largest might be.

Oldest tree A specimen of Bristlecone Pine (*Pinus aristata*, or *P. longaeva*) from California's White Mountains has been reliably dated at over 4800 years, by counting growth rings. Other woody plants have had greater ages attributed but they are ages of a whole colony, not of a single trunk.

Fastest-growing tree Several species from several countries contend for this record. A difficulty in comparing growth rates is that forest tree trials report different time intervals against which height growth is measured. Often the highest growth rates occur in plantations outside a species' native country. One that is hard to beat is Rose Gum (*Eucalyptus grandis*) in Hawaii, with 2-year-old trees 35 feet (10.7 m) high and 15-year-old trees 161 feet (49 m) high. The Moluccan legume Batai (*Paraserianthes falcataria*) was reported to reach 35 feet (10.7 m) in 14 months in Borneo and 107 feet (32.5 m) in 9 years. In southern USA the native Cottonwood (*Populus deltoides*) has grown to 43 feet (13 m) in 3 years, and to over 100 feet (30 m) in 9 years.

Most abundant tree This would most certainly be one of the dominant conifer species of the boreal forest (*taiga*) in subarctic Asia or North America, since these forests extend over vast areas and consist of very few species; but it is doubtful whether any comparative statistics exist. The strongest contenders are likely to be the North American White Spruce (*Picea glauca*) and the Siberian Larches (*Larix gmelinii and L. sibirica*).

Glossary

Abscission The shedding by a plant of leaves, flowers, fruits or other parts from a stem.

Alkaloid A chemical compound produced by a plant that in nature helps deter predation by insects or other herbivores due to its bad (often bitter) taste or by affecting the predator's physiology. Common plant alkaloids include caffeine, strychnine and cocaine.

Angiosperm A flowering plant. Angiosperms produce seeds contained within a fruit.

Apical Term indicating the location of a plant part or tissue at the growing tip (apex) of a shoot or root.

Aril A fleshy covering around or appendage to the seeds of some plants.

Auxin Any of the plant hormones that regulate growth and other processes.

Axil The upper angle between a leaf stalk (petiole) and the stem from which it grows.

Bark The surface layer of tissues that wraps around the trunk or branches of a tree or shrub and protects the wood underneath. Bark develops from a thin layer of dividing cells (cambium) and may become thicker as a tree grows. Some tree species shed patches or all of their bark each year.

Berry A fleshy, usually juicy fruit, such as a grape, tomato or watermelon, which has seeds embedded in its tissues.

Binomial The two-part scientific name of a plant species, such as *Acer rubrum* (Red Maple). The first part denotes a genus and the second part identifies the particular species within that genus.

Bipinnate Term for a type of compound leaf that has leaflets further subdivided into leaflets, so overall it is "doubly pinnate."

Bract A modified leaf that commonly protects a flower or inflorescence branch in bud.

Bud The embryonic stage of a leafy shoot, often protected by hard or leathery bud scales; a young, developing flower before its petals have expanded.

Calyx The outermost, usually green whorl of a flower, made up of the flower's sepals.

Cambium A thin layer of dividing cells in stems and roots that gives rise to the secondary growth that results in increasing girth of a woody plant. Wood (secondary xylem) is produced from its inner side, and bark from its outer.

Carpel The female reproductive organ of flowering plants. Depending on the species, flowers may have one, several, or many carpels, each enclosing one or more ovules that will develop into seeds.

Catkin A long, often pendulous inflorescence that generally includes many small, petalless flowers of only one sex. Trees such as oaks, alders and willows produce catkins.

Cell The smallest, most basic unit of life. Plant cells consist of a protoplast surrounded by a cell wall.

Chlorophyll The pigment in leaves and stems of green plants that captures sunlight during photosynthesis.

Common name The non-scientific name for a plant species, such as Red Maple or Crape Myrtle. Because common names vary greatly from country to country or region to region, botanists generally use scientific names to describe plants.

Compound term used to describe a leaf or other plant organ that consists of smaller segments. For example, the blade of a compound leaf is divided into two or more leaflets.

Cone In most conifers and all cycads, a reproductive structure that consists of scales arranged around a central axis. Seeds develop on seed (female) cone scales, while pollen grains develop in small sacs on scales of pollen (male) cones. Some conifers have highly modified, fleshy cones, hardly recognizable as such.

Conifer Any of the cone-bearing pinophytes, such as pines, firs, cypresses and yews. Conifers do not produce flowers or fruits but have naked seeds borne on the female cone scales. All are woody trees or shrubs and many have needle-like leaves.

Corolla The collective name for the petals of a single flower. The corolla is usually the most distinctively colored, showiest part of a flower.

Cotyledon The first leaf formed from a germinating seed, also called a seed leaf. Monocot plants have one cotyledon and dicot plants generally have two.

Cultivar A unique plant variety that exists as a result of cultivation or breeding for agricultural or ornamental use. Cultivar names are distinguished from species or subspecies names in being mostly non-Latin, capitalized and enclosed in single quotes, such as *Malus* 'Golden Hornet'.

Cuticle The protective waxy coating on the aboveground parts of a plant. The cuticle consists of wax and a fatty substance called cutin.

Cycad Any of a group of palmlike gymnosperms represented today mainly by species in tropical and subtropical habitats. Cycad leaves are long, mostly pinnate fronds radiating from a short trunk.

Deciduous Term for trees that shed all their leaves, in fall or in a tropical dry season, then regrow them in the spring or in the wet season.

Dicotyledons or dicots One of the two main groups into which flowering plants have traditionally been divided. Distinguishing features include two seed leaves (cotyledons), petals and sepals commonly in fours or fives, and leaves with netlike veins. However molecular and other studies now show this is not a natural group and that some dicot plant families should now be grouped with the "basal angiosperms," the remainder being distinguished as the "eudicots."

Dioecious Having male and female reproductive parts on different individuals of the same species.

Drupe A fleshy fruit that has one or more seeds enclosed in a hard inner layer or "stone." Stone fruits such as plums are examples of drupes.

Endemic As used for a species or genus, restricted to a particular geographic area or region.

Essential oil A concentrated oily liquid with a distinctive scent typical of the plant species. Essential oils usually are distilled from seeds, leaves, wood, bark or other plant parts. Eucalyptus oil and peppermint oil are examples.

Evergreen A tree or other plant that does not shed its leaves at the end of the growing season.

Family The taxonomic group below the rank of order that contains one or more genera. The names of plant families end in *-aceae*, as in the pine family Pinaceae.

Flower The reproductive structure of flowering plants, the angiosperms. Flowers vary greatly depending on the species but all contain at least one carpel (female) or one stamen (male).

Frond A large, divided leaf, such as the leaves of many ferns, palms or cycads.

Fruit In flowering plants, a mature, ripened ovary or group of ovaries containing the seeds.

Genotype The sum total of all genes present in an individual.

Genus The taxonomic group (plural, genera) below the rank of family which contains one or more species. For example, the various species of oaks are all in the genus *Quercus*.

Graft A piece of stem from one plant (the scion) that is joined with the rooted stem of a different plant (the rootstock).

Gymnosperm A non-flowering plant that produces naked seeds, not enclosed in a fruit. Conifers (Pinophytes) and cycads (Cycadophytes) are the major groups of gymnosperms.

Heartwood Wood in the center of a tree trunk in which the water-transporting vessels have become blocked and non-functional. Heartwood often is darker than sapwood.

Herb A plant that does not form a woody stem, such as a violet or daffodil. Herbaceous species generally are smaller than woody shrubs and trees.

Hybrid A plant that develops from the cross pollination of two different species and carries a combination of genes from the parent species.

Inflorescence Any specialized flower-bearing branch of a plant. There are various types of inflorescences, with the flowers occurring in a definite pattern in each type.

Lateral A shoot that grows from the side of a stem. Laterals develop from the axils of leaves.

Leaflet Term for the leaflike segments of a compound leaf.

Legume Any plant belonging to the large pea family, Fabaceae. Acacias are examples of leguminous trees.

Lignin A strengthening polymer in the cell wall of many vascular plants. Lignin's mechanical strength is a key feature in allowing trees to grow as tall as they do.

Mid-rib The large central vein of a leaf.

Monocotyledons or monocots One of the two major groups into which flowering plants have traditionally been divided (*see also Dicotyledons*). Their distinguishing features include a single seed leaf, flower parts in multiples of three and leaves with parallel veins.

Monoecious Term applied to species in which individual plants have both male and female reproductive organs.

Mycorrhiza A fungus that has a symbiotic association with the roots of a plant. Most vascular plants have mycorrhizae, which help supply the plant with crucial nutrients such as nitrogen. In return the fungal partner is nourished by sugars the plant manufactures in photosynthesis.

Nectar A sugar-rich fluid produced by glands in the flowers or on other parts of some plant species.

Node The place on a stem where one or more leaves attach.

Nutrient A substance such as nitrogen that is required for the normal development and growth of a plant.

Ovary The enlarged base of a carpel (or a set of fused carpels) that contains the ovules and matures into a fruit.

Ovule The structure in a seed plant containing a female sex cell (egg). After pollination and fertilization, the ovule develops into a seed.

Palm Any of the more than 2500 species in the flowering plant family Arecaceae. Most palms have large divided leaves and occur in tropical or subtropical regions.

Perennial A plant that lives for more than two years.

Perianth Botanical name for the structure formed by the combined petals and sepals of a flower.

Petal One of the parts of a flower. Petals collectively make up the corolla and often are brightly colored.

Phenotype The physical traits of an organism which result from its genes (genotype) interacting with its growing environment.

Phloem One of the two types of vascular tissue in plants. Phloem vessels mainly carry sugars produced in photosynthesis.

Photosynthesis The process by which green plants convert energy in sunlight to chemical energy, in the form of sugars.

Pinnate A compound leaf with feather-like arrangement of leaflets along both sides of a common axis or rachis.

Pollen Collective term for pollen grains, the sporelike structures produced by seed plants that contain male sex cells.

Pollination The transfer of pollen to female reproductive parts.

Pome A fruit formed by expansion of a flower's receptacle to enclose the developing ovary. Pears and apples are pomes.

Prickle A small, sharp-pointed extension from the epidermis of a plant stem or leaf.

Ray (or medullary ray) A pipeline-like array of cells that transport water and dissolved minerals laterally through the wood of a tree trunk.

Resin A sticky aromatic fluid produced by trees, notably the conifers, that is thought to protect against attack by certain insects and fungi.

Root Plant organ that usually grows through the soil, anchors the plant and absorbs water and mineral nutrients. Some plants, including many orchids and mangroves, develop aerial roots.

Rootstock The stump of a tree or vine that retains a healthy root system onto which a cutting from another plant can be grafted.

Sap The name given to the fluid contained in a plant's vascular tissues (xylem and phloem).

Sapwood The outer layer of wood in a tree trunk. It is usually lighter in color than the heartwood and contains all the functioning water-conducting vessels.

Scale leaf A specialized leaf that protects a dormant bud, or any of the small, overlapping leaves of conifers, such as cypresses.

Scion A branch or bud from a desirable tree or shrub that is grafted onto the rootstock of a different plant.

Seed The mature ovule of a seed plant, which contains a plant embryo. By means of seeds, which may be dispersed away from a parent plant by the wind, animals or other means, flowering plants and gymnosperms give rise to new generations.

Sepals The outer, often leaflike parts of a flower that typically enclose the other flower parts in the bud.

Shoot Generally, the aboveground part of a plant including stems, leaves, buds and other organs. The term may also be used to refer to any actively growing stem or branch.

Shrub A woody plant of smaller stature than a tree, often branching into many stems from ground level.

Simple leaf A leaf that consists of a single blade.

Species A level of classification of organisms, usually defined as comprising populations within and beween which there are no genetic barriers to continued interbreeding. The name of a species is a binomial, such as *Homo sapiens* or *Pinus sylvestris*.

Stamen The basic male flower reproductive structure consisting of a pollen-containing anther attached to a stalk or filament.

Stigma The site on the upper surface of a carpel where pollen grains adhere and germinate. With germination a channel called a pollen tube grows down into the carpel and delivers the genetic material of the grain to the egg inside the ovule.

Stilt root A type of aerial root in the form a prop at the base of a tree's trunk.

Stomate (plural stomata) A minute pore in a leaf surface that can open and close to allow oxygen, carbon dioxide and water vapor to pass inward or outward.

Subsoil The layer of soil beneath the topsoil. Usually the subsoil contains less organic matter and is of poor texture, but may be richer in the mineral nutrients required for plant growth.

Subspecies The main taxonomic grouping below the species level. Botanists further divide plant subspecies into varieties.

Succulent A plant with fleshy stems or leaves that store water.

Symbiosis A relationship in which (usually) two dissimilar organisms, such as a tree and a fungus, live in close association. Some symbioses are beneficial to both organisms while in others one partner is a parasite.

Tannin An astringent chemical produced by some plants that helps defend buds, seeds, bark and other tissues against insect predators and other threats.

Taxonomy The subfield of biology that studies the classification of organisms.

Terminal A bud or other part, such as a flower, that is located at the tip of a shoot.

Thorn A short, woody, sharp-pointed branch.

Transpiration The loss of water vapor from leaves and other aerial plant parts.

Variety A level of classification of plants below the species level. Some botanists equate varieties with subspecies, or alternatively divide subspecies into varieties.

Vascular plant A plant having an internal system of specialized cells that transport water, sugars and other substances throughout the plant body.

Vegetative A term used to describe plant parts not involved in reproduction, such as leaves; in vegetative reproduction a plant is propagated by asexual means such as by cuttings.

Vein A bundle of transportation vessels that forms part of the framework of a leaf or some other organ.

Venation The pattern of veins in a leaf.

Wood The strong, hard secondary xylem tissue formed by and inside the cambium layer of a tree, adding girth to its trunk.

Woody plant A plant that forms woody tissues, generally during a lifespan of many years.

Xylem One of the two types of conducting tissues in plants. Xylem transports water absorbed by roots upward throughout the plant body.

Index

A

acacias 21, 68, 77, 78, 79, 84, 210, 211
acetylsalicylic acid 250
acid rain 270
    causes and effects 271
acorn weevil 193
Adansonia 116
aerial roots 47
African Blackwood 240
African Oil Palm 120
African rain sticks 240
agricultural uses for wood 242
airplanes 233
alders 86, 184
Aleppo Pine 254
Alerce 20, 148, 187
algarrobas 211
almond 67
Amazon River 27
amber 238, 239
American Aspen 29, 37
American black bear 184
American brown pelican 216
American Calabash Tree 242
American Elm 174
angiosperms 24, 60, 64, 93, 108, 110, 229
annatto 254
annual rings 40, 74
    cross-section 75
anthers 56, 58, 60, 66
anthracite 252, 253
apical dominance 70
apples 66, 68, 93, 98
aquatic trees 124–5
    mangroves 125
    soil, in 125
Araucaria 106, 186
Archaeanthus 60
Arctic bison 185
Arctic tundra 182
Arizona mountain king snake 201
ash trees 87, 191, 241
aspens 86
Athel Tree 171
attack and defense 76–9
    chemical defenses 77
Australian eucalypt forests 194, 196, 197
    wildfires 194, 197, 199
Australian Mountain Ash 126, 145
Australian Overland Telegraph Line 233
Australian Soft Tree Fern 97
autumn 53
avalanches 86

B

Babul 127
bacteria 78
badgers 219
Bago 150
Baja Elephant Tree 158
Bald Cypress 124, 171
balm of Gilead 250
Balsa 157
balsa wood 229
balsam fir 250
bamboo 244
banana plants 18
Banda Aceh, Indonesia 85
Banyan 47, 147
Baobab 37, 63, 129
    trunks 37
bark 34, 36, 40, 44
    decorative 44, 45
bark huts 230
Barnsley House, Gloucestershire 258
barrelmaking 243
basswood 191, 239
bats 57, 62
Battle of N·jera 236–7
Beach Casuarina 137
Beach Pandanus 158
beeches 18, 83, 117, 194, 238
    germination process 116
beer 254
berries 66
biodiversity 272–3
biomes 180, 202
birches 86, 184
birds 57, 62, 189, 193, 197, 201
bituminous coal 252, 253
Bixa orellana 254
bixin 254, 255
Black Bean Tree 251
Black Birch 44
Black Spruce 254
Black Tree Fern 97, 142
Black Walnut 152
Black Wattle 127
Blackjack Oak 95
Blackwater Falls State Park, Virginia 190–1
blowpipe 236
Blue Atlas Cedar 105
Blue Gum 194
Blue Quandong 239
Blue Spruce 186
boats 234–5
    figureheads 238
bobcat 200
Bodhi Tree 224
Bois de Boulogne, Paris 218
boiserie 231
bonsai 260, 261
Boojum Tree 19, 148
boomerang 236
boreal forest 27, 106, 180, 182–5
    animals and birds in 184–5
    conditions in 183
    dominant trees in 184
    locations 182
boreal owl 185
Bornean orangutan 205
botanical gardens 260
botanical hierarchy 25
Bottle Tree 37
bracket fungus 193
branches and twigs 34, 38–9, 42
    branch modifications 39
    branching patterns 39
Brazil nuts 67
breadfruit 249
breeding 261
Bridgeman, Charles 258
bridges 232
Bristlecone Pine 72, 105, 161, 186
broadleaf deciduous forests 180, 190–3
    clearing 190
    climate 190
    life in 193
    location 190
    productivity and diversity 191, 192
    winter 193
broadleaf trees 184
    temperate broadleaf deciduous forests 180, 190–3
    temperate southern broadleaf forests 194–7
    temperate woodlands 199
Brown, Lancelot ``Capability” 258
brown lemur 208

Bull Bay Magnolia 111, 260
Bunya Bunya 64, 132
buttress roots 46

C

cabinetmaking 238
  Chippendale 239
Cacao 66, 119, 173, 248
cacti 122–3
Calabash Tree 141
California condor 200
California Red Fir 126
California Incense Cedar 243
California Redwood 44
calyx 60
cambium 40
camphor 239
Camphor Laurel 110
Candelabra Tree 123, 146
Cannonball Tree 62, 141
canoes 234, 235
canopy 38, 202, 204
Capability Brown 258
carbon cycle 267
carbon dioxide 226, 266, 267, 268
carbon trading 280
Carboniferous period 252
cardboard 255
caribou 185
carpel 66
carpentry 228
carvings 231, 239
  Polynesian figural wood carvings 239
castanospermine 251
caterpillars 78
Cathedral Rock, USA 198–9
cauliflorous blooming 63
Cecropia 267
Cedar of Lebanon 137
cedars 98, 106, 186
Ceibo Trees 206–7
cellophane 255
cellulose fibers 244, 255
Central Park, New York 218, 219, 256
cerrado, Brazil 211
chameleon 208–9
charcoal 242, 253
chariots 232
checkers 241
chemical defenses 77
Chernobyl Nuclear Power Plant disaster 271
cherry 66, 238
cherry blossom 256
chess 241
chestnut 67, 68
chicle 248
Chinese Camellia 135
Chinese papermaking techniques 244
Chippendale furniture 238, 239
chlorophyll 35, 50, 52, 54, 55
chloroplasts 35, 55
chocolate 66
Cholla 123
chopsticks 242
Christmas trees 224, 254
Church of the Transfiguration, Russia 230
cicadas 49, 196
Cinchona tree 250, 251
Cinnamon Tree 45, 139, 248, 249
citrus fruits 248
cladodes 39, 51, 53
classification of plants as trees 18
classification of tree vegetation 180
classification of trees 92–3
climate 26–9
  extremes 272
  ice storms 29
  impact of 28, 180
  leaves and 52
  microclimate 26
  temperate woodlands and 199
climatic variables 180
  refugia 186
Clove 170
coaches and carts 232–3
coal 252–3
coal mines 252–3
Coast Redwood 20, 21, 103, 168
coastal regions 86
coco de mer 68, 69
cocoa 67, 248
coconut 67
Coconut Palm 82, 113, 120, 121, 140, 247
cocopeat 247
coffee 248, 249
coir 247
competition from other plants 80, 81
cones 64–5, 103, 104
  fertilization 64, 65
  types of 64
coniferous forests
  distribution of 186
  life in 188–9
  succession in 280
conifers 18, 74, 75, 86, 87, 102–7, 184, 226, 229
  amber 238
  characteristics 102
  needles 102
  shape 70, 71, 82
  temperate woodlands and 199
  wind and weather, surviving 82
conkers 241
conservation 272–3
  sustainable development principles 272
construction
  homes and shelters 230–1
cooking 253
coral islet 27
Corcovardo Mountain, Brazil 218
cork 246
Cork Oak 45, 165, 246
corolla 60
Cottonwood 164
coyotes 219
crab apples 109, 267
Crabapple Mangrove 214
Crassulacean Acid Metabolism 122
cricket bats 240, 241
customs 224
cycads 18, 24, 60, 100–1
  distribution 100
  fronds 100, 101
  genera 100
  sago palm 101
  structure 100
cypresses 104, 186

D

Date Palm 113, 120, 121, 159
Dawn Redwood 155
deciduous trees 38, 52, 94–5, 184, 256
  climate and 29
  leaves 94
  temperate broadleaf deciduous forest 180, 190–3
  temperate woodlands and 199

tropical dry deciduous forest 206–9
decorative arts 238–9
decorative woods 231
deer 193, 200
defense and attack 78–9
definition of tree 18
deforestation 274, 278–9
causes of 279
effects of 279
dehiscent 68
dhows 235
dicotyledons (dicots) 18, 74, 92, 93, 98, 110, 112, 118
digeridoo 240
dipterocarp 206
disease 78
DNA 35
Douglas Fir 51, 104, 164, 186
Dove Tree 143
Dragon Tree 52, 113, 144
drink 248–9
drums 241
drupes 66, 67
dry deciduous forest 180, 206–9
conditions 206
food chain 208
locations 206
treelife in 206
wet and dry seasons 208
dryad's saddle fungus 78
dugout canoes 234, 235
Durian 66, 144
Dutch East India Company 249
dwarf willow 184
dyes, thinners and tannins 254

E

early plant life 22
earthworms 49
Eastern chipmunks 193
Eastern White Pine 163
eating utensils 242
Ebony 43, 236, 238, 240
Ecuador 63
El Arbol del Tule 21
elephants 78, 79
elk 185
Elkhorn Cedar 104
elms 19, 118, 191
energy and fuel 252–3
English Oak 42, 73, 165
Enkoji Temple, Japan 256–7
epiphytes 36, 188, 194, 204, 206
erosion reduction 266
espalier 260, 261
eucalyptus 20, 44, 84, 85
bark 44, 45
blossoms 58, 69
leaves 77
oil 77
seeds 68, 69
eucalyptus oil 250
Euphorbia 18, 122
European Beech 146
European Larch 38
European Spruce 186
evergreens 52, 94–5, 224, 256
evolutionary chart of plants 23
extinction, threat of 272

F

fabrics 255
fall coloring 191
fallow deer 193
false fruits 66
families of trees, related 118
Faurea 194
ferns 87, 96–7, 188
"crozier" stage 97
locations 96
vascular structure 96
fertilization and reproduction 56–7
cross-fertilization 56
tropical trees 116
fiber products 246–7
fibrous root system 46
Ficus macrophylla 46
figs 47, 66, 79, 147, 206
figureheads 238
fire 84–5, 279, 281
firs 183, 184, 186, 228
fishermen's houses 231
Flamboyante 143
flood 84–5
protection 276–7
Floss Silk Tree 76, 139
flowering and non-flowering plants 60
Flowering Dogwood 59
flowering trees 108–19
insects and 108
lifecycle 109
flowers 58–63
dioecious 58
evolution 60
fertilization 60
fragrance 63
hermaphroditic 58
monoecious 58
pollination 34, 62
reproductive organs 57
shapes 60
structure 60
flutes 240
fly agaric 188
foliage
deciduous 94
evergreen 94
wind and weather, surviving 82
food 248–9
colorings and flavorings 254
forest clearing 275, 277
forest tent caterpillar 78
forestry 226
forests 180–1
boreal 180–5
broadleaf deciduous 190–3
coniferous 186–9
distribution map 181
major forest biomes 180
microclimates 26
old growth 226
prehistoric 22
reforestation 280–1
Slovenia 180–1
sustainable 281
temperate southern hemisphere broadleaf 194–7
tropical dry deciduous 206–9
tropical rain forest 27, 180, 202–5
water conservation 276
woodlands 198–201
fossil fuels 252, 269
fossils 98
foxes 219
fragrance 63

Frangipani 61, 119
Frankincense 134
Franklin Tree 149
French Alps 83
French polishing 238
freshwater mangrove 255
fruit 66–7, 68
  Ginkgo 99
fruit trees 248
fungi 49, 78, 188, 192, 196
furniture 238
  Chippendale 239

G

galleons 234
galls 78
games 241
gardens 218, 256–61
Gardens of Versailles 256–7
General Sherman *Sequoiadendron*, USA 126
genetics 62, 70
geographical range 28
germplasm 68
giant mud crab 217
Giant Sequoia 75, 168, 186
Ginkgo 18, 22, 24, 98, 149, 250, 251, 261
  seeds 99
ginkgo nuts 67
giraffe 213
global warming 182, 268
goanna 196
Golden Hind 234
golden poplars 267
Gondwana 194
Grass Tree 21, 175
gray mangroves 214
Great Green Wall, China 281
greenhouse effect 268–9
greenhouse gases 269
greenhouses 260
grevillea 68
grizzly bear 184, 185
grosbeaks 188
ground-cones 188
growth 34, 74–5
  knots 43
  stages 72–3
  survival of the fittest 80–1
  variation 42
gum mastic 250
gymnosperms 24, 60, 64, 67, 92, 102

H

hail 82
Hanging Gardens of Babylon 256
hardwoods 43, 226, 228–9
Hawaii 87
Hawaiian Tree Fern 97
hazel 236, 253
hazelnut 67, 68
heartwood 40, 42, 74
heating 252–3
henna dye 254
Hermanophyton 22
Herrania balaensis 62, 63
hesperidium 66
Heyerdahl, Thor 234
hibernation 193
hickory 241
hickory chips 255
hierarchy of classification 25
hollies 94
homes and shelters 230–1
Honey Locust 86
Horse Chestnut 130
  conkers 241
Horseradish Tree 156
humus 48, 49
Huon Pine 153, 235
hurricane 82

I

ichneumon wasp 189
indehiscent 68
Indian Rhododendron 166
inflorescence 60
inlay timbers 238
inner city wildlife 219
insects 78, 108, 208
  pollination 56, 59, 62, 63
Iriartea deltoidea 121
ironwood 236
Italian Alder 56

J

Jacaranda 53, 116, 151
Jack Pine 87
Jackalberry Tree 212
Japanese Maple 21, 108, 128
Japanese Umbrella Pine 167
Japanese Yew 242
Jardin du Luxembourg, Paris 259
Java 87
Jefferson Memorial, Washington DC 256
Jeffrey Pine 228
Joshua Tree 39, 123
Judas Tree 61, 138
Jumping Cholla 123
jungle 206
junipers 64, 103
Juniperus osteosperma 21

K

Kaempfer, Engelbert 99
kakapo 197
kangaroos 196
kapok 247
Kapok Tree 46, 138
Kauri 130, 126, 186
Kent, William 258
Kigelia spp. 62
Kilmarnock Willow 62, 63
King Billy Pine 235
Kitty Hawk 232
knots in timber 228
koala 196
Kon-Tiki 234
Krakatau Island 87
Kudzu Vine 81

L

Laburnum 61, 258
Lacebark Pine 161
lamina 50
land management practices 274
landscape garden movement 258
landscaping 256–61
  suburban 218
landslips 86, 87

larches 27, 64, 65, 94, 183, 184, 186
American larches 182
latex 246, 247
Laurel Tree 224
leaf color 99, 126, 191
leaf litter 49
leafcutter ants 205
leaves 34, 50–5
adaptations 51
climate and 52
color 52
deciduous 94–5
diversity 51
Ginkgo 99
photosynthesis 54, 94
stomata 54
wet climates, in 53
legume genera 211
lemons 248, 249
lemurs 63
leopard 213
lichens 188
lightning 83
lignite 252, 253
lignophytes 23, 100
limes 109, 239, 249
Linnaeus (Carl von Linnè) 24, 122
Loblolly Pine 107
logging 226–7, 278
longships 236
lumber 229
lumberjacks 226
lupines 87
lycophytes 22

M

macadamia 67
mace 236
machines of war 236–7
Madagascan palm 246
magnesium 48
magnolia 60, 61, 69
Bull Bay Magnolia 111
Mahogany 169, 238
major tree groupings 92
mallee growth 70
Man Fern 97
Manchurian Cherry 45
mangoes 68
mangroves 47, 86, 125, 180, 214–17, 218, 277
conditions 214
fish 216
flood protection 276
food chain 217
life in 216
locations 214
major genera of 214
root systems 125
manipulation, art of 160–1
Manzanita tree 44
maples 117, 191, 200, 240, 241
bark 44
seeds 68
sugar 254
syrup 248
Maritime Pine 254
marsupials s196
Mastic Tree 250
maturity 72
MDF (medium density fiberboard) 244
medicine 250–1
Mediterranean Cypress 142
Melaleuca alternifolia 250
mesquites 211
metamorphic rock 252
microclimate 26
Mignonette Tree 254
mimosas 211
minerals 48
mining 275, 2778
mistletoe 79, 188
mitochondria 35
moisture 46
monarch butterflies 188
monilophytes 23
Monkey Puzzle 37, 39, 51, 94, 131
monocotyledons (monocots) 110, 112
seed 92
Monterey Pine 162, 226
Montezuma Cypress 21, 172
moose 185
moraines 86
Moreton Bay Chestnut 251
"mosaic burning" 85
mosses 188
Mount St Helens, Washington 87
Mountain Ash 87
mountain lion 201
mudflats 86, 87
mudskippers 216
Mugo Pine 254
mulberry 66, 218, 244
music and entertainment 240–1
mythology 224–5

N

Nageia minor 124
naming trees 24–5, 99
Native American tools 243
necrophytes 78
nectar 58
Neem 133, 251
nematodes 78
New Zealand Mamaku 97
New Zealand Soft Tree Fern 97
nitrogen 48
non-flowering plants 60
Norfolk Island Pine 239
North American porcupine 185
Norway Spruce 38, 254
Nothofagus antarctica 194
Nothofagus forests 194, 196
Nutmeg Tree 156
nutrients 46
nuts 67, 249

O

oaks 84, 118, 191, 192, 234, 238, 246, 260
barrels for wine 242
evergreen 200
leaves 51
shape 70
oil refining 271
Olive tree 52, 157, 248, 249
Olmstead, Frederick 256
Ombú 160
orange 66, 67
Organ-pipe Cactus 122
Oriental Plane 163
ornament and amenity 256–61
ornamental trees 257
owls 185, 193
Oyamel Fir 188

P

pachycaul 19
Pacific Madrone 132
Pacific Yew 251
palms 18, 82, 112, 120–1, 211
- by-products 120
- cross-section of trunk 121
- landscaping, in 259
- shape 70, 71
- trunks 37, 82

paper 244–5
Paper Birch 133
Paper Mulberry 134
paperbarks 44, 250
papermaking 244–5
- early 244
- modern 245

papyrus 244
parasitic plants 79, 80, 81, 188
parenchyma 42
parks 256
Patas monkeys 213
pawpaw 111
peach tree 109
peanut 67
peat 252, 253
pecan 255
Pecharo-Ilychsky Nature Reserve 183
peg roots 47
pencils 243
perennial plants 18
pericarp 66, 67, 68
permafrost 183
pests 78
petals 58, 60, 66
petiole 50
Philippines 87
phloem 40, 41, 44
phoenix palms 259
phosphorus 48
photosynthesis 34, 35, 50, 54, 94, 191, 268
phyllodes 39, 51, 53
pignut 255
pigs, wild 200
pine marten 185
pine needle oil 254
pine nuts 67
pines 64, 65, 83, 84, 87, 104, 105, 183, 184, 186, 200, 228, 238
pinesaps 188
Pinites succinifera 239
pinyons 199
pioneers 86–7
pistillate flowers 58
plane tree 78
plant cells 35
plantations 226
plants
- classification as trees 18
- evolutionary chart 23

pleaching 258
plows 242
plums 66, 67, 68
plywood 228
pneumatophores 47, 125
podocarps 106, 124, 186
Pohutukawa 155
pollarding 260, 261
pollen 56, 58, 60, 98, 109, 110, 116
- pollution 270

pollination 34, 59, 62–3
- vectors of 57

pollinators 53, 57, 60, 62, 205
pollution 270–1
- acid rain 270–1
- nuclear 271

Ponderosa Pine 162, 254
poplars 184, 199, 267
possums 196, 219
Princess Tree 159
proboscis monkey 216
pruning 260–1
pseudogalls 213
pteridophytes 22
pulp and paper 244–5
- environmental effects of industry 244

pulp mills 244

Q

quinine 250, 251
Quiver Tree 29, 113, 131

R

raccoons 219
raffia 246, 247
raffia palm 120, 247
rafts 234
rain forests 180, 202–5
rainfall 26, 82, 180, 208, 277
Rambutan 117
recycling 244
red-eyed leaf frog 205
Red Mangrove 166, 214–15
red squirrel 189
redwoods 186
reforestation 280–1
refugia 186
reproduction 56–7, 58
resins and fragrant woods 77, 239, 250
Rhus Tree 77
Rio de Janeiro 218
River Red Gum 124, 145
rocking horses 241
role of trees 266–7
roots 34, 46–7, 48
- aboveground 47
- cross-section 47
- structure 46
- wind and weather, surviving 82

rosewood 228, 240
rosin 254
Rowan Tree 87
rowans 184
rubber 246–7
- products 247

Rubber Tree 151, 246
Russian dolls 241

S

Sacred Fig 52, 225
sacred trees 225
sago palm 101
Saguaro 19, 122, 136
- cross-section 122

Sal 169
salicin 250
sand and gravel bars 86
sand dunes 275
sandalwood 239
Sandplain Cypress 64
sapwood 40, 41, 42, 74
Sausage Tree 62

savanna 21, 180, 210–13
  animals 211, 212
  conditions 210
  dry season 213
  locations 210
  plant life 211
  termites 212
sawmills 226–7, 229
scent 63
schizocarpic 68
sclerenchyma cells 41
sclerophylly 51
scrub growth 70
scurvy 249
seasons 52–3
seeds 68–9
  biodiversity protection 272
  conifers 106
  dispersal 86–7
  fire germination 68, 84
  formation 68
  fruit 66, 68
  Ginkgo 99
  structure 68
self-pollination 62
sepals 60, 66
sequestration 268
Serengeti Plain, Tanzania 210–11
Shagbark Hickory 136, 255
shape 70–1
shields 236
ships 235
Shortleaf Pine 45
siege towers 236
Silky Oak 150
Sitka Spruce 160, 186
size of trees 20–1
skittles 241
skunks 219
slavery 249
sleds 232
smaller trees 21
Snake-bark Maple 44
snow 82
Snow Gums 119
snowplant 188
softwoods 43, 226, 228–9
soil 48–9
  life in 49
  stabilization 274–5
soil erosion 226, 278
  control of 274–5
southern hemisphere broadleaf forest 194–7
  life in 196–7
  locations 194
Southern Magnolia 154
Species Plantarum 24
spices 248, 249
spiders 208
spines 76, 77
spring 52
"Spruce Goose" 233
spruces 64, 86, 107, 183, 184, 186, 228, 240, 254
  Norway 82, 83
squirrels 185, 189, 193, 218, 219
stamens 60, 66
staminate flowers 58
star anise 248
Stick on Line 241
stigma 58, 59, 60
stilt houses 231
stilt roots 47
stomata 54
Stradivari, Antonio 241
Strangler Fig 79, 80, 117, 147
stringed instruments 240
strip cutting 273
structure 34–5
subsoil 48
suburbs, trees in 218–19
succulents 122–3
sugar glider 197
Sugar Maple 95, 126, 129, 191
summer 52
sustainable development 272
  strip cutting 273
sustainable forests 281
Swamp Cypress 47
Swollen Thorn Acacia 117
sycamore 56, 240
symbolism 224–5

T

Tabebuia 116
taiga 106, 182, 183
Talipot Palm 140
Tamarack 183
Tamarind 170
tarantula 208
Tasmanian Blue Gum 250
Taxodium mucronatum 126
taxol 251
taxonomy 24
tea tree oil 250
Teak 173
temperate forest 49
  broadleaf deciduous 180, 190–3
  coniferous 180, 186–9
  southern hemisphere broadleaf 180, 194–7
  woodlands 180, 198–201
temperatures 180, 182, 190, 268
Terminalia 211
termites 212
testa 68
thorns 76
three-toed sloth 267
Tibetan Cherry 45
Tibouchina 119
Tijuca Forest 218
timber, trees for 226–9
timber categories 228
timber paneling 231
timber production 226–9
  reforestation 280
toadstools 188
toco toucan 204
tools and utensils 242–3
  modern 243
topiary 260, 261
topsoil 48
tornado 83
totem poles 224
tracheids 43
train tracks 233
transpiration 34, 51, 95
transport 232–3
  water, across 234–5
trebuchet 237
tree ferns 18, 34 *see also* ferns
Tree of Knowledge, the 225
Tree of Life, the 224
tree shapes 70–1
trees, uses for 228–59
  decorative arts 238–9
  energy 252–3

food 248–9
homes and shelters 230–1
medicine 250–1
music and entertainment 240–1
ornamental 256–61
pulp and paper 244–5
rubber, cork and fiber 246–7
timber 226–9
tools and utensils 242–3
travel and exploration 232–5
trees for flowers, foliage and bark 257
Trojan Horse 236
tropical Batai 36
tropical forest
dry deciduous 180, 206–9
rain 94, 180, 202–5
Tropical House, Royal Botanic Gardens, Kew 260
Tropical Paperbark 154
tropical rain forest 27, 94, 180, 202–5
diversity of life-forms 202
interdependent relationships 205
life in 204–5
regeneration 202
structure 202
succession 202
true fruits 66
trunks 34, 36–7
cross-section 40
eyes 39
scarring 39
spiny 76
wind and weather, surviving 82
tsunami 85
Tucson, Arizona 218
Tulip Tree 58, 111, 153, 191
tundra 182
tupelo 239
tyloses 42
types of trees 92–3

U

Umbrella Acacias 211
Umbrella Thorn 20, 71, 128
urban landscape 218–19
inner city wildlife 219
Utah Juniper 152

V

Valonia Oak 191
Vaux, Calvert 256
vegetation, classification of 180
veins of leaves 50
Versailles 256
vines 79
violins 240, 241
viruses 78
volcanic eruptions 86, 87

W

wallabies 196
walnut 67, 191, 228, 236
warships 236
wasps
fig 66
ichneumon 189
water conservation 276–7
wattles 53, 254
waxes for wood 238
weapons 236–7
weather 82–3
wedge-tailed eagle 197
weeds 79
Weeping Willow 70, 71, 167
Western Hemlock 226
Western Red Cedar
Whistling Thorn 213
White Oak 191
wildfires 194, 197, 199
willows 118, 184, 199, 241
Kilmarnock 62, 63
Weeping 70, 71, 167
white 250, 251
wind 27, 82, 180
wind instruments 240
winged samara 68
winter 53
Witch Hazel 63
Wollemi Pine 103, 126, 174, 273
wolves 201
wombat 196
wood 18, 40–3, 74
types of 40
wood chips 227, 244
wood creosote 254
wood-drying kiln 229
wood pulp 244
wood veneer 227, 228
woodlands 21, 180–1
climate 199
conditions 198
juniper-pinyon 199, 200
large mammals of 201
locations 198
pine 27
riparian 199
temperate 198–201
woodpeckers 185
woodworking tools 242
World Tree 224

X

xylem 41, 42

Y

Yangtze River, flooding 277
Yellow Birch 191
Yew 21, 172
Yggdrasil 224
Ylang Ylang 63, 135
yuccas 18

Credits

PHOTOGRAPHS

t=top; l=left; r=right; tl=top left; tcl=top center left; tc=top center; tcr=top center right; tr=top right; cl=center left; c=center; cr=center right; b=bottom; bl=bottom left; bcl=bottom center left; bc=bottom center; bcr=bottom center right; br=bottom right

AAP=Australian Associated Press; AUS=Auscape International; CBT=Corbis; GI=Getty Images; iS=istockphoto.com; MP=Minden Pictures; NPL=naturepl.com; PL=photolibrary.com; SH=Shutterstock; USGS=United States Geographical Survey

Front cover tl iS; tc, GI; tr PL; bl GI; br CBT
Back cover and spine Peter Bull Art Studio
1c CBT; **2**c MP; **4**r GI; **6**c MP; **8**c CBT; **10**c, l CBT; r GI; **11**c, l CBT; cr PL; **12**c PL; **14**c CBT; **16**tcl, tcr PL; **17**c GI; cl CBT; l PL; **18**bl, c PL; **19**br, tr PL; **20**r PL; **21**b, tl, tr PL; **22**bl PL; **24**bl PL; br SH; tr GI; **25**c, cl, cr, l, t, tc, tcl, tcr, tr SH; **26**bcr GI; tr MP; **27**bl PL; tl GI; **28**b PL; **29**b CBT; tr MP; **30**c CBT; **32**c, l, r PL; **33**l PL; tc SH; **34**bl SH; **35**bcl, bcr PL; t PL; **36**r PL; **37**bc SH; br, tc PL; tl CBT; **38**bl SH; br, tr PL; **39**bl, tl PL; tr SH; **40**c PL; **41**cr PL; tr PL; **42**b GI; **43**br SH; tl PL; **45**bc SH; bl, br PL; **46**br, tr PL; **47**r GI; **48**b PL; tr PL; **49**l CBT; r PL; **50**r SH; **51**bc PL; tc, tr SH; tl MP; **52**c PL; **53**bcr, br PL; tr PL; **54**bl PL; **55**b GI; **56**bcr, r PL; **57**bc PL; bl, br SH; **58**b PL; **59**bl, tl SH; tr PL; **60**b, t PL; **61**b, t SH; **62**bl PL; c PL; **63**br, tl PL; **64**bl AUS; br, tr SH; **65**bcr PL; r PL; **66**bl, tr SH; **67**bc, tr SH; bl, br, tl PL; **68**bl PL; r SH; **69**br iS; l, r, tr SH; **70**br PL; tr GI; **71**bl, tl GI; **72**br SH; tr PL; **73**bl, tc SH; tl PL; **75**br PL; c, tr CBT; **76**r SH; **77**bl SH; tl PL; tr PL; **78**bl PL; br PL; **79**b, tl PL; **81**b CBT; t SH; **82**b PL; **83**bc, tl PL; tr SH; **84**r PL; tr PL; **85**l PL; r, tcr PL; **86**b MP; **87**br CBT; tr MP; **88**c GI; **90**bcr, l SH; c MP; cl PL; **91**c, tcl PL; **92**r GI; **93**l, r GI; **94**b MP; **95**bl SH; br PL; tl MP; **96**bl SH; br AUS; **97**bcl PL; br Forest and Kim Starr tl GI; **98**r CBT; **99**br PL; tl GI; tr CBT; **100**bl SH; bl AUS; tr PL; **101**c SH; **102**c MP; tr SH; **103**br MP; tl AUS; **104**bl MP; tr PL; **105**bcr PL; tr MP; **106**bl MP; tr SH; **107**bl MP; br SH; t GI; **108**b GI; **109**bl, tr GI; br SH; **110**bl, br PL; **111**b, tl, tr PL; **112**r CBT; **113**br, l PL; tr CBT; **114**bl MP; **115**br GI; c, tr MP; **116**r SH; **117**br SH; tl MP; tr GI; **118**br, l SH; **119**br PL; tl MP; **120**tr CBT; tr GI; **121**br PL; t GI; **122**tr MP; **123**b MP; t PL; **124**bl CBT; r AUS; **125**tl AUS; tr PL; **126**br CBT; **127**br AUS; tr USGS; **128**br SH; tr PL; **129**br PL; tr AUS; **130**br, tr PL; **131**br AUS; tr PL; **132**br PL; tr CBT; **133**br SH; **134**br, tr PL; **135**br PL; tr TR **136**br, tr PL; **137**br PL; tr USGS; **138**br PL; tr TR **139**br PL; tr PL; **140**br PL; tr SH; **141**br PL; tr PL; **142**br PL; tr SH; **143**br, tr PL; **144**br, tr PL; **145**br, tr AUS; **146**br PL; tr SH; **147**br CBT; tr USGS; **148**br, tr CBT; **149**br SH; tr PL; **150**br USGS; tr Donovan Bailey, New Mexico State University **151**br, tr SH; **152**br SH; tr PL; **153**br PL; tr AUS; **154**br AUS; tr PL; **155**br AUS; br PL; **156**br PL; tr TR **157**br PL; tr GI; **158**br USGS; tr CBT; **159**br, tr PL; **160**br CBT; tr Gerald Carr **161**br PL; tr AUS; **162**br USGS; tr MP; **163**br, tr PL; **164**br, tr SH; **165**br SH; tr PL; **166**br PL; tr AUS; **167**br PL; tr SH; **168**br PL; tr AUS; **169**br PL; tr CBT; **170**br PL; tr CBT; **171**br PL; tr John Stretch, Department of Agriculture and Food WA **172**br SH; tr CBT; **173**tr GI; tr SH; **174**br TR tr PL; **175**tr AUS; **176**CBT; **177**c CBT; **178**l, tc MP; tcr PL; **179**c, l MP; **180**b GI; **182**c PL; **183**tr PL; **184**bl SH; r MP; **185**tl, tr PL; **186**bl PL; **187**c PL; **188**b AUS; **189**bl, tl MP; br PL; **190**c MP; **191**tr PL; **192**b MP; **193**br MP; tl, tr PL; **194**bc PL; **195**c GI; **196**bl AUS; br SH; tr MP; **197**bc MP; t PL; **198**c MP; **199**tr GI; **200**b MP; **201**bl, br MP; tl Anup Shah/NPL; **203**c MP; **204**bl SH; br MP; **205**br, tr MP; **206**bl CBT; **207**c CBT; **208**bl SH; tr PL; **209**c MP; **210**c PL; **211**tcr MP; **212**b MP; tcr PL; **213**br, tr MP; tl Larry Michael/NPL; **214**bl PL; **215**c AUS; **216**bl, tr PL; br MP; **217**b MP; **218**bl MP; br PL; **219**r PL; **220**c CBT; **222**cl PL; cl SH; **223**l PL; tc CBT; **224**bl, br CBT; tr PL; **225**r SH; tcl PL; **226**r GI; **227**bl, l PL; br, tl GI; **228**bc, bl, br SH; tr PL; **229**bc, bl, br PL; **230**b SH; tcr PL; **231**b, tcr PL; c, tc GI; **232**cr SH; tr GI; tr PL; **233**br PL; tr CBT; **234**bl AAP; **235**cr, tl PL; **236**b, c CBT; t GI; **237**c GI; **238**bl, br PL; **239**b, tl PL; tr CBT; **240**bcl, br GI; bl, tr PL; **241**b CBT; tc PL; **242**br, tr GI; **243**b CBT; bl, tl GI; tr SH; **244**bl, tr PL; **245**bl, br PL; **246**bl PL; tr CBT; **247**bl CBT; r PL; tr SH; **248**b, tr SH; **249**bl, br PL; cr, tr SH; **250**c PL; **251**b, tr PL; c, tcl PL; **252**bl, br PL; **253**tl, tr PL; **254**r PL; **255**bl, tr PL; br AUS; tl SH; **256**bl PL; **257**b CBT; t PL; **258**l PL; **259**bl, br, tl PL; **260**b CBT; **261**br CBT; tc, tl PL; tr iS; **262**c PL; **264**cl GI; l CBT; tc MP; tr PL; **265**c CBT; **266**b SH; **267**bc SH; c CBT; tr PL; **268**bl SH; r GI; **269**b SH; **270**b GI; **271**tl, tr PL; **272**bcr MP; bl AAP; **273**br PL; t GI; **274**bl PL; r PL; **275**r, tr MP; **276**bl CBT; **277**bl, br GI; t PL; **278**cr MP; **279**bl, cl MP; tr GI; **280**tr PL; **281**b PL; tc CBT; tr SH; **282**c GI; **284**bl, SH

ILLUSTRATIONS

Darren Awuah/The Art Agency 102bl, 104tc, 109tl, 116tl, 120bc
Peter Bull Art Studio 34tr, 40l, 43bl, 45tr, 47l, 54tr, 55tr, 57tr, 72tc, 74c, 77r, 80b, 92bcl, 96tr, 100r, 114b, 122b, 185bc, 189tr, 202cr, 217tr, 229tr, 245t, 246br, 267tl, 284c, 285b c t tl, 286b c t
Leonello Calvetti 222cr, 234cr
Andrew Davies Creative Communication 23c, 25c
Andrew Davies Creative Communication and Map Illustrations 181t, 182bl, 186c, 190bl, 194cl, 198bl, 202bl, 206cl, 210bl, 214c, 287b, bc, bcl, bcr, bl, br, c, cr, t, tcl
GODD.com 237b
Steven Hobbs 273bl
James McKinnon 226l
John Francis/ Bernard Thornton Artists UK 22tr
Michael Saunders 253b
Terry Pastor/The Art Agency 125br, 271bl, 280b
Steve Trevaskis 235c

SILHOUETTES

Andrew Davies Creative Communication

Captions
page 1 This Buddha head in Phra Nakhon Si Ayutthaya, Thailand, has been enfolded by the roots of a Banyan Tree.
page 2 Baobab trees store water in their swollen trunks. These examples are from Madagascar.
page 4–5 For a few short weeks each year, the Jefferson National Forest, Virginia, USA, is ablaze with the colors of fall.
page 6–7 A grove of Birch Trees at dawn, Northwoods, Minnesota, USA.
page 8–9 Trees cling to the banks of a river, surrounded by mudfats in the Kakadu National Park, Australia.
page 12–13 Ferns grow in the understory of a temperate rain forest in the Olympic National Park, Washington State, USA.
page 282–3 A forest lake in Sichuan Province, Peoples Republic of China.

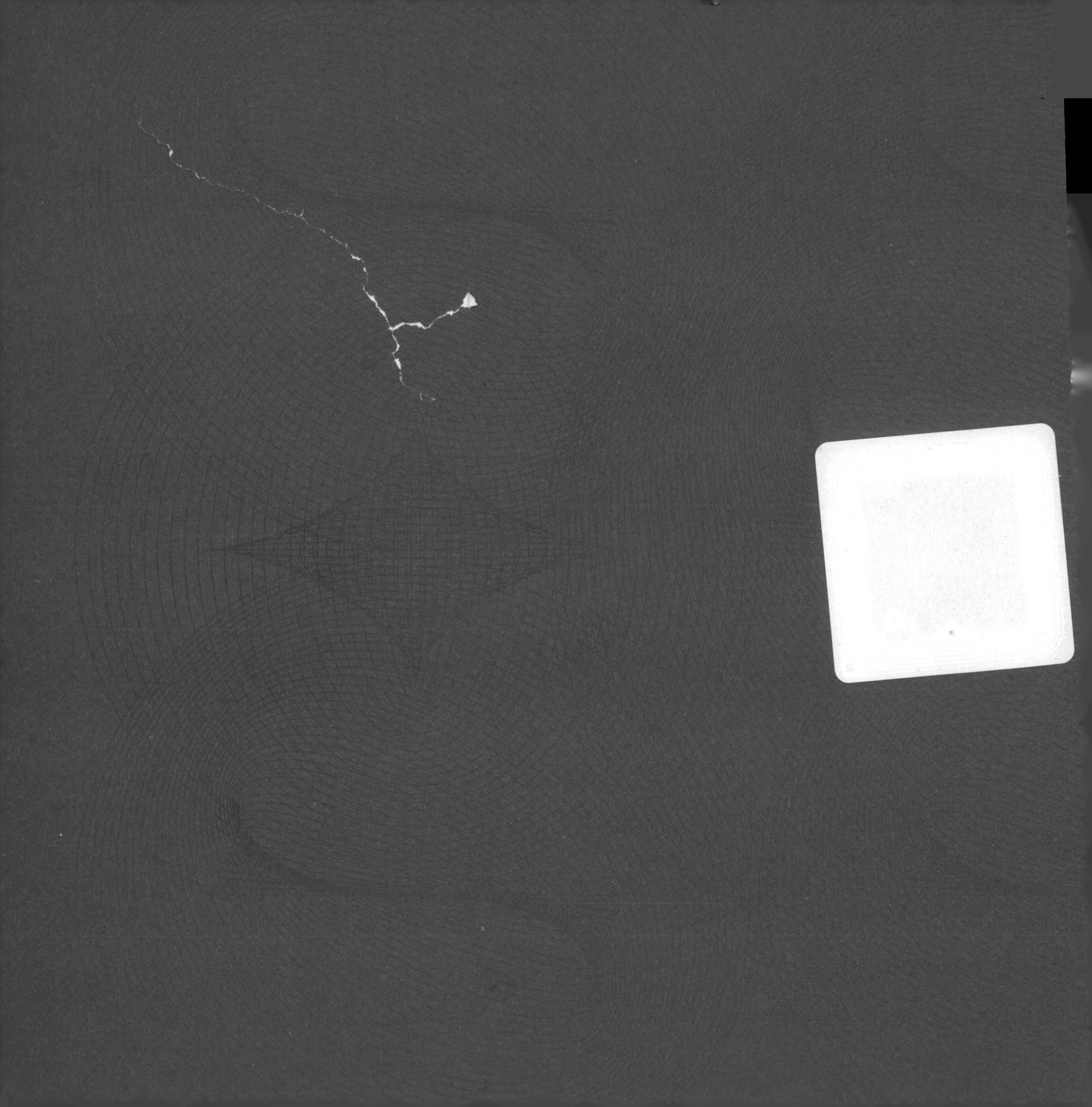